GREEN HEALTHCARE INSTITUTIONS
Health, Environment, and Economics
WORKSHOP SUMMARY

Howard Frumkin and Christine Coussens, *Rapporteurs*

Roundtable on Environmental Health Sciences, Research, and Medicine

Board on Population Health and Public Health Practice

INSTITUTE OF MEDICINE
OF THE NATIONAL ACADEMIES

THE NATIONAL ACADEMIES PRESS
Washington, D.C.
www.nap.edu

THE NATIONAL ACADEMIES PRESS • **500 FIFTH STREET, N.W.** • **Washington, DC 20001**

NOTICE: The project that is the subject of this report was approved by the Governing Board of the National Research Council, whose members are drawn from the councils of the National Academy of Sciences, the National Academy of Engineering, and the Institute of Medicine. The members of the committee responsible for the report were chosen for their special competences and with regard for appropriate balance.

This study was supported by contracts between the National Academy of Sciences and the National Institute of Environmental Health Sciences, National Institutes of Health (Contract N01-OD-4-2193, TO#43); National Center for Environmental Health and the Agency for Toxic Substances and Disease Registry, Centers for Disease Control and Prevention (Contract No. 200-2000-00629, TO#7); National Institute for Occupational Safety and Health, Centers for Disease Control and Prevention (Contract 0000166930); National Health and Environment Effects Research Laboratory and the National Center for Environmental Research, U.S. Environmental Protection Agency (Contract 282-99-0045, TO#5); American Chemistry Council (unnumbered grant); ExxonMobil Corporation (unnumbered grant); and Institute of Public Health and Water Research (unnumbered grant). The views presented in this book are those of the individual presenters and are not necessarily those of the funding agencies or the Institute of Medicine.

Additional copies of this report are available for sale from the National Academies Press, 500 Fifth Street, N.W., Box 285, Washington, DC 20055. Call (800) 624-6242 or (202) 334-3313 (in the Washington metropolitan area), Internet, http://www.nap.edu. For more information about the Institute of Medicine, visit the IOM home page at **www. iom.edu.**

The serpent has been a symbol of long life, healing, and knowledge among almost all cultures and religions since the beginning of recorded history. The serpent adopted as a logotype by the Institute of Medicine is a relief carving from ancient Greece, now held by the Staatliche Museen in Berlin.

Cover photo: Reprinted with permission, from Guenther 5 Architects, PLLC. Copyright (2005) by David Allee.

Suggested citation: Institute of Medicine (IOM). 2007. *Green healthcare institutions; Health, environment, and economics (Workshop summary)*. Washington, DC: The National Academies Press.

*"Knowing is not enough; we must apply.
Willing is not enough; we must do."*
—Goethe

INSTITUTE OF MEDICINE
OF THE NATIONAL ACADEMIES

Advising the Nation. Improving Health.

THE NATIONAL ACADEMIES
Advisers to the Nation on Science, Engineering, and Medicine

The **National Academy of Sciences** is a private, nonprofit, self-perpetuating society of distinguished scholars engaged in scientific and engineering research, dedicated to the furtherance of science and technology and to their use for the general welfare. Upon the authority of the charter granted to it by the Congress in 1863, the Academy has a mandate that requires it to advise the federal government on scientific and technical matters. Dr. Ralph J. Cicerone is president of the National Academy of Sciences.

The **National Academy of Engineering** was established in 1964, under the charter of the National Academy of Sciences, as a parallel organization of outstanding engineers. It is autonomous in its administration and in the selection of its members, sharing with the National Academy of Sciences the responsibility for advising the federal government. The National Academy of Engineering also sponsors engineering programs aimed at meeting national needs, encourages education and research, and recognizes the superior achievements of engineers. Dr. Wm. A. Wulf is president of the National Academy of Engineering.

The **Institute of Medicine** was established in 1970 by the National Academy of Sciences to secure the services of eminent members of appropriate professions in the examination of policy matters pertaining to the health of the public. The Institute acts under the responsibility given to the National Academy of Sciences by its congressional charter to be an adviser to the federal government and, upon its own initiative, to identify issues of medical care, research, and education. Dr. Harvey V. Fineberg is president of the Institute of Medicine.

The **National Research Council** was organized by the National Academy of Sciences in 1916 to associate the broad community of science and technology with the Academy's purposes of furthering knowledge and advising the federal government. Functioning in accordance with general policies determined by the Academy, the Council has become the principal operating agency of both the National Academy of Sciences and the National Academy of Engineering in providing services to the government, the public, and the scientific and engineering communities. The Council is administered jointly by both Academies and the Institute of Medicine. Dr. Ralph J. Cicerone and Dr. Wm. A. Wulf are chair and vice chair, respectively, of the National Research Council.

www.national-academies.org

ROUNDTABLE ON ENVIRONMENTAL HEALTH SCIENCES, RESEARCH, AND MEDICINE*

Paul Grant Rogers (*Chair*), Partner, Hogan & Hartson, Washington, D.C.

Lynn Goldman (*Vice Chair*), Professor, Bloomberg School of Public Health, Johns Hopkins University, Baltimore, Maryland

Jacqueline Agnew, Professor, Bloomberg School of Public Health, Johns Hopkins University, Baltimore, Maryland

John Balbus, Director of Health Program, Environmental Defense, Washington, D.C.

Roger Bulger, Advisor to the Director, National Center on Minority Health and Health Disparities, Bethesda, Maryland

Yank D. Coble, Immediate Past President, World Medical Association, Neptune Beach, Florida

Henry Falk, Director, Coordinating Center for Environmental and Occupational Health and Injury Prevention, National Center for Environmental Health/Agency for Toxic Substances and Disease Registry, CDC, Atlanta, Georgia

Baruch Fischhoff, Howard Heinz University Professor, Department of Engineering and Public Policy, Carnegie Mellon University, Pittsburgh, Pennsylvania

John Froines, Professor and Director, Center for Occupational and Environmental Health, Southern California Particle Center and Supersite, University of California, Los Angeles

Howard Frumkin, Director, National Center for Environmental Health/ Agency for Toxic Substances and Disease Registry, CDC, Atlanta, Georgia

Paul Glover, Director General, Safe Environments Programme, Health Canada, Ottawa, Ontario

Bernard Goldstein, Professor, Department of Environmental and Occupational Health, Graduate School of Public Health, University of Pittsburgh, Pennsylvania

Myron Harrison, Senior Health Advisor, ExxonMobil, Inc., Irving, Texas

Carol Henry, Acting Vice President for Industry Performance Programs, American Chemistry Council, Arlington, Virginia

John Howard, Director, National Institute of Occupational Safety and Health, Washington, D.C.

Peter Illig, Consultant, Association Internationale pour l'Ostéosynthèse Dynamique, Trauma Care Institute, Nice, France

Richard Jackson, Adjunct Professor, Environmental Health Services Division, University of California at Berkeley

---

*Membership current as of December 31, 2006.

Reviewers

This report has been reviewed in draft form by individuals chosen for their diverse perspectives and technical expertise, in accordance with procedures approved by the National Research Council's Report Review Committee. The purpose of this independent review is to provide candid and critical comments that will assist the institution in making its published report as sound as possible and to ensure that the report meets institutional standards for objectivity, evidence, and responsiveness to the study charge. The review comments and draft manuscript remain confidential to protect the integrity of the deliberative process. We wish to thank the following individuals for their review of this report:

Ellen Dorsey, Program Officer, Heinz Endowments, Pittsburgh, PA
Paul R. Fisette, Director, Building Materials and Wood Technology, Holdsworth Natural Resources Center, University of Massachusetts, Amherst
Lt. Gen. Henry J. Hatch (Ret), Former Chief of Engineers, United States Army, Oakton, VA
Susan West Marmagas, Director of Health Programs, Collaborative on Health and the Environment, Blacksburg, VA

Although the reviewers listed above have provided many constructive comments and suggestions, they did not see the final draft of the report before its release. The review of this report was overseen by **Ada Sue Hinshaw,** Professor, University of Michigan School of Nursing, who was responsible for making certain that an independent examination of this report was carried out in accordance with institutional procedures and that all review comments were carefully considered. Responsibility for the final content of this report rests entirely with the authoring committee and the institution.

Contents

Preface

This workshop is the ninth in a series of workshops sponsored by the Roundtable on Environmental Health Sciences, Research, and Medicine since the roundtable began meeting in 1998. When choosing workshops and activities, the roundtable looks for areas of mutual concern and also areas that need further research to develop a strong environmental science background.

When the Roundtable on Environmental Health Sciences, Research, and Medicine began its discussions, the roundtable members suggested that a broader concept of environmental public health needed to be established. The roundtable has built on other definitions of environmental health to include the natural, built, and social environments. Prior to these initial discussions, many roundtable members felt that there had been a focus on the toxicological effects of individual environmental agents to the detriment of understanding the larger picture of how environmental conditions impact health. The roundtable members acknowledged that the built environment—where and how communities and transportation systems are built—is very important and relevant to health.

The roundtable's first workshop, *Rebuilding the Unity of Health and the Environment: A New Vision of Environmental Health for the 21st Century*, examined and explored a broader definition of the environment as a very integral part of health. Where people live, work, and play impacts their health, and environmental policies must consider this relationship.

In the past six years, the role of the built environment has received more attention in public health. Roundtable members Howard Frumkin, Lynn Goldman, Richard Jackson, Samuel Wilson, and others helped to shape the roundtable's thinking on the built environment, bringing this issue to the forefront in public health leadership and planning communities.

This workshop focused on the environmental and health impacts related to the design, construction, and operation of healthcare facilities, which are part of one of the largest service industries in the United States. Healthcare institutions are major employers with a considerable role in the community, and it is important to analyze this significant industry. The environment of healthcare facilities is unique. It has multiple stakeholders on both sides, as the givers and the receivers of care. There are ill and injured individuals, their families and friends, and the employees that deliver care to them. Many of the most vulnerable individuals pass through the doors of healthcare facilities each day.

In order to provide optimal care, more research is needed to determine the impacts of the built environment on human health. The scientific evidence for embarking on a green building agenda is not complete, and at present, scientists have limited information. There is general information that pleasant places that emit low levels of chemical materials are good for the environment and good for health, but, at best, science can make only vague statements. For example, there is no guideline to determine how much use of natural daylight as a source of illumination is necessary to realize benefits. Overall, the major point that I took away from the workshop is that the scientific community needs to think strategically about its funding in this area. There is an opportunity of great promise, yet more information about the complexities involved in building a green facility is needed. A number of speakers pointed out that hospitals, which regularly collect information on patient outcomes, are ideal living laboratories to advance knowledge as the United States embarks on replacing many facilities from the early postwar era. Through implementation of controlled studies, investigators can address the research gaps and discern the complexities of building green on human health. The challenge will be to conduct meaningful research in this area that examines the interplay of the built environment and health. Finally, the workshop participants discussed research directions that will help promote an environment for overall health.

This workshop summary captures the discussions and presentations by the speakers and participants; they identified the areas in which additional research is needed, the processes by which change can occur, and the gaps in knowledge. The views expressed here do not necessarily reflect those of the Institute of Medicine, the roundtable, or their sponsors.

Paul G. Rogers, *Chair*
Roundtable on Environmental Health
Sciences, Research, and Medicine

1

Introduction[*][†]

Future historians of the late 20th and early 21st centuries may well mark the growth of environmentalism as one of the epochal transformations of the time. Governments, industries, and the public have come to understand the importance of sustainability and of environmental protection, and the necessary science, technology, and policy have evolved rapidly. The healthcare sector, which accounts for one-sixth of the U.S. economy, has come relatively late to environmental thinking, but the rise of "green health care" signals a major step forward.

Green health care—the incorporation of environmentally friendly practices into healthcare delivery—appeals to health professionals and institutions for many reasons. It offers the potential to safeguard the environment, an increasingly compelling challenge. It allows healthcare institutions to demonstrate leadership in their communities. It can be a platform for educating students and members of the public. It can save money. Each of these rationales was acknowledged at the workshop on green health care held by the Institute of Medicine (IOM) Roundtable on Environmental Health Sciences, Research, and Medicine on January 10–11, 2006. However, for health professionals like those represented at the workshop, green health care is likely to be most compelling because of its potential to protect and promote health, both directly and indirectly.

These health benefits may operate on at least three scales: local, community, and global. On the local scale, within the walls of a hospital, research facility, or clinic, green construction and operation can protect patients, workers, and visi-

---

*The planning committee's role was limited to planning the workshop, and the workshop summary has been prepared by the workshop rapporteurs as a factual summary of what occurred at the workshop.

†This chapter is an edited version of the opening remarks and the summation by Howard Frumkin at the workshop. The presentations were combined to eliminate duplication.

1

tors. For example, choosing safe cleaning agents or limiting the use of pesticides can reduce the potential for toxicity among those exposed. On the community scale, reducing the ecological footprint of a hospital reduces environmental hazards and protects natural resources. For example, linking a hospital to its community with pedestrian infrastructure and mass transit can reduce motor vehicle traffic and help achieve clean air. Reducing packaging in the hospital cafeteria or adopting biodegradable cutlery and plates can reduce the volume of waste sent to landfills. On the global scale, green practices help steward scarce resources and reduce environmental degradation. For example, a hospital that purchases food or supplies from local sources reduces the need for long-distance transport of goods, thereby reducing the associated greenhouse gas emissions that contribute to climate change. A hospital that installs flooring made from sustainably harvested wood helps slow deforestation, which in turn preserves biodiversity and the livelihoods of faraway rural populations.

WHAT IS GREEN HEALTH CARE?

Green building can be defined in many ways. According to the Office of the Federal Environmental Executive, "green or sustainable building is the practice of designing, constructing, operating, maintaining, and removing buildings in ways that conserve natural resources and reduce pollution" (OFEE, 2003). This definition is fully applicable to healthcare facilities at all stages of design, construction, and operation. *The Green Guide for Healthcare* (2006) identifies opportunities to enhance environmental performance in the following domains: site selection, water conservation, energy efficiency, recycled and renewable materials, low-emitting materials, alternative transportation, daylighting (the use of natural light in a space to reduce electric lighting and energy costs), reduced waste generation, local and organic food use, and green cleaning materials. Some decisions, such as site selection, occur during the planning and construction phases; other decisions, such as food sourcing and cleaning practices, are primarily questions of operation after a building is completed. Commitments to energy conservation, renewable resource use, and similar principles must be made and reinforced throughout the life cycle of a facility, from building conception through operation and replacement.

At the IOM roundtable workshop, an even broader perspective emerged. Participants discussed "natural" features of buildings, such as daylighting, gardens, and nature views, some of which may offer health benefits. They discussed functional aspects of hospital design, such as legibility, coherence, and way-finding cues, which are used to orient oneself within the built environment and may ease and humanize the experience of being a visitor or patient. They discussed health-promoting design features and messages, such as attractive staircases that lure people from using elevators. Characteristics such as these, taken together with green principles, offer a positive vision of sustainable, health-promoting healthcare settings.

In fact, in creating a vision that resonates with health professionals and leaders, "framing" the concept of healthcare facility design for public and environmental well-being is critical. Characteristics of such framing include:

- Aspirational: Green healthcare facilities aim not only to avoid harm, but also to enhance well-being and to restore the environment.
- Economical: Green healthcare facilities provide value and save money.
- Prudent: Green health care reduces future risks, such as those related to energy price shocks, building-related health problems, and building obsolescence.
- Long-term: Some benefits of green buildings emerge over years, or even over the entire life span of a building.
- Contextual: Green buildings yield benefits not only within their own walls, but also in the context of the community or even the national or global arena.

A constant theme of the workshop was complexity. Designing, constructing, and operating buildings require careful balancing of a vast array of variables. Careful analysis using systems thinking is essential. Craig Zimring of the Georgia Institute of Technology, during his presentation, warned of the "fallacy of generalized goodness"; not all green decisions are all good. For example, although wide hallways, large rooms, and oversize windows that provide natural daylighting may create pleasant environments for staff and patients, they may also increase energy demand and costs. The presence of plants may pose challenges for infection control. Thoughtful analysis, supported by empirical data and a culture of continuous improvement, is necessary.

Green healthcare principles can be implemented on many scales, from physicians' offices, clinics, and community hospitals to vast medical centers that occupy several city blocks. At the workshop, most discussion focused on large hospitals and academic medical centers, not only because these are the venues in which many Institute of Medicine (IOM) members work, but also because data are most plentiful from such settings. Moreover, large institutions offer strategic advantages: health science students are trained there, so effective green healthcare principles can be modeled and disseminated. Also, many large institutions are currently undertaking building programs, offering opportunities for far-reaching impact. Even so, participants noted that there is an important role for environmentally friendly practices at every level of the healthcare system.

WHY PURSUE GREEN HEALTH CARE?

In both the public sector and the private sector, the concept of the "triple bottom line" has become well established in recent years (Elkington, 1998; Esty and Winston, 2006; Savitz and Weber, 2006; Willard, 2002). This concept, sometimes summarized as "people, planet, and profit," holds that the best performance for a firm, agency, or institution is one that optimizes social, environmental, and economic outcomes. For healthcare institutions, the social dimension includes health impacts. Thus a hospital with a successful triple bottom line would boast positive impacts on the health and well-being of its patients, staff, and visitors; efficient use of energy and natural resources, with minimal waste and pollution generated; and healthy financial performance. Many private firms recognize that this approach not only advances their goals but also positions them well on the market, enhancing their image and earning customer loyalty. These rationales apply directly to green health care.

There are also ethical reasons for pursuing green health care. Biomedical ethics are usually based on four principles: autonomy, beneficence, nonmaleficence, and justice (Beauchamp, 2001; Engelhardt, 1995). The provision of green health care is especially consistent with beneficence, as it provides benefits to patients and staff (and, in a larger sense, to communities near and far and to unborn generations), and with nonmaleficence, as it avoids harms (including distant downstream harms) that could result from certain conventional practices. Public health ethics have been linked to three traditions—utilitarianism, liberalism, and communitarianism (Roberts and Reich, 2002)—and these also offer a compelling rationale for green health care. Utilitarians would point out that the net sum of human well-being—considering patients, staff, visitors, community members, and others—is likely to increase if healthcare institutions are green. Liberal analysts, following Kant, would argue that the right to a healthy environment is infringed by policies and practices that permit dangerous exposures. And communitarians would argue that the necessary conditions for "good society" are enhanced by green health care. *The Principles of the Ethical Practice of Public Health* begins with a statement that, prima facie, supports green health care: "Public health should address principally the fundamental causes of disease and requirements for health, aiming to prevent adverse health outcomes." (Thomas et al., 2002). Thus, green health care falls squarely in the traditions of both biomedical and public health ethics.

WHAT IS THE EVIDENCE SUPPORTING GREEN HEALTH CARE AND WHAT FURTHER EVIDENCE IS NEEDED?

What one does in health care—the medications administered, the surgical procedures performed, the health behaviors that are recommended to the public, the systems that are fashioned to deliver services—ought to be safe and effective, and these attributes ought to be established by evidence. In the case of therapies, the gold standard of evidence is the randomized clinical trial (Chow and Liu, 2004; Friedman et al., 1999; Katz, 2006). In the case of systems, performance-based measurement provides evidence that supports continuous improvement (Gawande, 2007; IOM, 2000; Smith, 2005) . With respect to economic outcomes, careful analysis of costs and benefits can provide the evidence base for wise decisions (Brent, 2003; Donaldson et al., 2002 ; Drummond et al., 1997).

Similarly, the move to green health care should be supported by evidence. Simply claiming that something is *green*, without demonstrating empirical benefits for human health and well-being, the environment, and economics, is not enough. Although anecdotal accounts of success and case studies are useful in advancing green health care, a robust evidentiary base for the practice is needed. Many endpoints might be studied—patient health outcomes, staff turnover, the psychological comfort of visitors, the magnitude of the waste stream, the use of water, and the cost of energy, to name a few.

> Simply claiming that something is *green*, without demonstrating empirical benefits for human health and well-being, the environment, and economics, is not enough.
>
> —*Howard Frumkin*

An important goal of the workshop was to discuss some of the lines of research that, if carried out, could help guide the transition toward green health care.

HOW IS GREEN HEALTH CARE IMPLEMENTED?

Healthcare institutions, like any complex systems, do not change easily or quickly. Many factors drive change, including the evolution of new external demands, the emergence of new data, the reframing of questions about a healthcare institution's operation, the influence of visionary leadership, and the reconciliation of competing interests within the institution. In the case of green health care, each of these may play a role; indeed each may be indispensable. Advocates of green health care need to understand the institutional dynamics, including the strengths and weaknesses of particular institutions and the opportunities and threats both at the institutional level and in the larger operating environment. Green healthcare success stories are useful in illuminating what works and what can go wrong.

2

Sustainable Healthcare Facilities

In 2004 the healthcare industry in the United States constituted 16 percent of the country's gross domestic product. Today, the United States is experiencing one of the largest healthcare facility building booms since the Hospital Survey and Construction Act, commonly referred to as the Hill-Burton Act, was passed in 1946 (Public Law 79-725). New technologies and new competitive pressures affect health care as more people are moving to the suburbs and older city hospitals are becoming obsolete. Hospital design transformation requires looking for new ways to improve healthcare quality and recognizing the relationship between medical services, environment, and diseases. Visionary thinking, connecting sustainability to health, and pollution prevention are important for the future of the healthcare industry.

The first session of the workshop is summarized in this chapter and includes presentations by Craig Zimring, Robin Guenther, and Knut Bergsland. This portion of the workshop explored the agendas involved in green health care and provided examples of sustainable healthcare facilities and the methods and tools employed in their design and operation. The information provided is drawn from the insights and experience of the presenters, and some of the current standards and best practices being implemented in green healthcare institutions today.

GREEN BUILDING AND HEALTH AGENDAS: POINTS OF CONVERGENCE

Because modern healthcare facilities are large consumers of resources, they provide an opportunity to make changes or reduce the consumption of these resources. Despite financial pressures, they are dedicated to helping people and can be models for other institutions, said Craig Zimring of the Georgia Institute of Technology. According to Zimring, there are two agendas for green health care: the green agenda and the design-as-quality-support (DQS) agenda. The green agenda is a multilevel analysis of socioeconomic health impacts at multiple scales, ranging from a building's occupants to society as a whole. The DQS agenda converges with the green agenda but differs, according to Zimring, in some important ways. Similar to the green agenda, the DQS agenda advances social and economic goals, but it focuses on using design to improve quality and safety outcomes, such as error and infection reduction, staff turnover, length of stay, and patient and family satisfaction. In this agenda, design is viewed as a tool that can affect healthcare outcomes for patients, staff, and the institution as a whole.

Relationship Between the Two Agendas

The green agenda is based on the notion of a virtuous cycle explained Zimring. Designing, constructing, and managing a hospital in accordance with principles of sustainable development can benefit the local community, the economy, and the environment. It can improve public health as well as reduce the demand for health services.

In contrast, the DQS agenda approaches construction from a slightly different perspective. Similar to evidence-based medicine, in which healthcare decisions can be made based on the best evidence about the outcome of those decisions, evidence-based design decisions are based on the best predictions about their outcome, asserted Zimring. Evidence-based design is the conscientious, explicit, and judicious use of current best evidence in making design decisions that advance an organization's goals. DQS involves a process in which one understands the evidence, makes hypotheses, tests the outcomes, and works back into decision making.

Part of the DQS agenda is based on two Institute of Medicine (IOM) reports: *To Err Is Human: Building a Safer Health System* (IOM 2000) and *Crossing the Quality Chasm: A New Health System for the 21st Century* (IOM 1999). In *To Err Is Human*, the IOM extrapolated that as many as 98,000 people die each year from preventable medical errors. Furthermore, there are approximately 2 million hospital-acquired infections in the United States, and as many as 88,000 people die from those infections (CDC, 2000). Zimring observed that even at the lower end of these numbers, more people in the United States die from hospital-acquired preventable adverse events than from other leading causes of death, such as AIDS, breast cancer, and motor vehicle accidents (IOM, 2000).

One of the reasons why these safety problems occur is that the most experienced caregivers in hospitals—nurses—have a very high turnover rate, noted Zimring. Concern about safety, quality, and nursing turnover has contributed to a "quality revolution" and the belief that healthcare institutions can make a dramatic improvement in healthcare quality and safety through better information and strategic action. This initiative is led by such organizations as the Institute for Healthcare Improvement, the Center for Health Design, and others.

Role of the Physical Environment in Green Building and Health

Zimring observed that there is a large and growing body of evidence demonstrating the role of the physical environment in achieving healthcare quality and safety. For example, a recent meta-analysis of more than 600 primarily peer-reviewed studies found associations between the physical environment and patient and staff outcomes in four areas: reduced staff stress and fatigue and increased effectiveness in delivering care; improved patient safety; reduced patient stress and improved health outcomes; and improved overall healthcare quality (Ulrich et al., 2004). For example, access to views and natural light in healthcare facilities can have important stress-reducing effects, as well as reduce pain and the length of stay at the hospital (Ulrich, 1991). In their meta-analysis review, the authors observe that hospitals are complex systems in which it is difficult to isolate the impacts of individual factors and suggest that design-based evidence parallels evidence-based medicine for improving health care. Zimring recommended additional research in the area of evidence-based design to understand how hospital design affects health.

In conclusion, he described the green agenda as encompassing ecological health on multiple scales. It credits environmental initiatives, such as reduced resource use, and it aims for improved patient outcomes. The green agenda does not guarantee that recycled floor surfaces or less outgassing of chemicals from products will result in better care or more rapid recovery, cautioned Zimring. In reinventing hospitals and transforming their design, design is a tool to improve quality, safety, and experience.

HIGH-PERFORMANCE HEALING ENVIRONMENTS

As construction technology advanced during the past century, the design of hospitals changed from daylit, naturally ventilated, pavilion-style buildings to high-rise buildings with mechanically conditioned air, said Robin Guenther of Guenther 5 Architects. Bellevue Hospital in New York City is a prime example of this change. It was built on a site near nature, overlooking the East River, before the expertise was available to build high-rise buildings or provide mechanical ventilation. At that time, hospitals needed access to clean air and fresh water to heal people. The facility has continued to expand and based on blueprint calculations, Guenther reported that Bellevue Hospital now has 60,000 square feet of floor space, an acre and a half footprint, and less than 10 percent of the building has a window.

Because projections for construction in the healthcare sector are for approximately 100 million square feet per year, Guenther suggested that architects need to identify new models for healthcare facilities. Today's healthcare facilities are buildings that are low in thermal mass—they heat up too quickly and cool down too rapidly—and are heavily dependent on artificial systems, such as lighting,

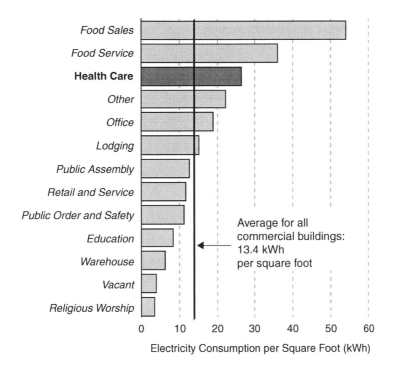

FIGURE 2-1 Healthcare facilities annually consume nearly twice the energy of an average commercial building.
SOURCE: Energy Information Administration, 1995 Commercial Buildings Energy Consumption Survey.

ventilation, and interior environments. Healthcare facilities consume nearly twice the annual energy of an average commercial office building (Figure 2-1).[*]

Even with healthcare expenditures constituting roughly 16 percent of the GDP, healthcare facilities are unable to keep up with demand and upgrade older facilities. Many of the structures built in the United States during the Hill-Burton Act era, particularly urban medical centers, are challenged by real estate and capital issues and are in need of replacement, said Guenther.

---

[*]The data presented by the speaker was from a 1995 report from the EIA. The report was updated in 2003 and shows that the average electricity consumption per square foot (kWh) for all commercial buildings is 14.1, and for healthcare facilities is 22.9.

Leadership in Energy and Environmental Design

In the past two decades, the link between buildings and health has received considerable attention. Many media reports have highlighted the problem of indoor air toxins in sealed buildings. The October 2003 issue of *Metropolis* magazine was entitled "Architects Pollute," and placed a large share of the responsibility for contributing to global warming on the shoulders of architects. An article within the issue asserted that architects and the construction industry are responsible for half of America's energy consumption and half of its greenhouse gas emissions produced by burning coal, gasoline, and other fossil fuels (Hawthorne, 2003).

Americans' growing concern that buildings *do* impact their health has influenced the federal government's decision to support green building. In its report *The Federal Commitment to Green Building: Experiences and Expectations,* green buildings are defined as "the practice of (1) increasing the efficiency with which buildings and their sites use energy, water, and materials, and (2) reducing building impacts on human health and the environment through better siting, design, construction, operation, maintenance, and removal—the complete building life cycle" (OFEE, 2003).

The U.S. Green Building Council developed Leadership in Energy and Environmental Design (LEED) as "a nationally accepted benchmark for the design, construction, and operation of high performance green buildings. The LEED rating system is the current best practice standard for the building sector. LEED gives building owners and operators the tools they need to have an immediate and measurable impact on their buildings' performance. LEED promotes a whole-building approach to sustainability by recognizing performance in five key areas of human and environmental health: sustainable site development, water savings, energy efficiency, materials selection, and indoor environmental quality" (U.S. Green Building Council, 2006).

Initially, LEED was created as a tool for the for-profit sector, which needed motivation to use emerging green building technologies, and was based on the premise that the nonprofit sector would embrace green building technologies as a component of their mission. LEED certification is given to the top 25 percent of green-performing buildings in the United States. With levels that include Certified, Silver, Gold, and Platinum, certification is achieved based on a third-party checklist of strategies for building, design, and construction. Of the approximately 1,800 registered projects in the LEED system in 2004, only 2 percent of these projects were in the healthcare industry (Kozlowski, 2004).

Examples of LEED Construction in Health Care

A few noteworthy LEED projects in health care include Emory University's Winship Cancer Institute in Atlanta, Georgia; the Dell Children's Medical Center in Austin, Texas; Boulder Community Foothills Hospital in Boulder, Colorado; and the Patrick H. Dollard Health Center in Harris, New York.

Emory University's Winship Cancer Institute is a research and clinical facility for cancer patients that formally received certification in the U.S. Green Building Council's LEED program in 2005. Input from patients influenced the design of the Infusion Center, which has more intimate clusters of space for four patients each and hip-high walls that personalize the space for family members while maintaining visibility for good nursing care. The building design seeks to create an atmosphere of life, health, and the ultimate hope for a cure.

Dell Children's Medical Center in Austin, Texas, is a LEED-registered project that is pursuing certification at the Platinum level. A redevelopment located at the old Austin Municipal Airport, it is pursuing green strategies across the board; most notably, the integration of a combined heat and power facility that doubles the efficiency of electricity generation by utilizing waste heat for thermal energy. It is scheduled for occupancy in late 2007.

Boulder Community Foothills Hospital is a replacement women's and children's hospital. It is a green-field hospital built on previously undeveloped, degraded wetlands that were home to a large prairie dog colony. The hospital returned 32 restored acres of the 49-acre site to the community as permanent open space. The city of Boulder then built the hospital sustainably as a way to gain community support for development of the site, explained Guenther. The hospital architects focused on local and regional low-emitting materials, reduction of water use through xeriscaping (landscaping that does not require supplemental irrigation), low-flow fixtures, and energy reduction.

Another sustainable healthcare facility is the Patrick H. Dollard Discovery Health Center in upstate New York. It is a residential facility for developmentally disabled children and adults, and the medical staff has a keen interest in the impact of the environment on developmental disabilities. The Dollard Health Center construction strategies included high-reflectance roofing, reliance on local and regional recycled-content low-emitting materials, and elimination of polyvinyl chloride in finish materials and plumbing. By using ground source heat pumps, energy demand was reduced by 42 percent and onsite fossil fuel combustion was eliminated. In addition, the facility opted to purchase green power, said Guenther.

Green Guide for Health Care

First released in 2003, *The Green Guide for Health Care* filled a need in the marketplace for green building tools specifically for health care. According to its website, *The Green Guide for Health Care* is "the healthcare sector's first quantifiable sustainable design toolkit integrating enhanced environmental and health principles and practices into the planning, design, construction, operations and maintenance of their facilities. This guide provides the healthcare sector with a voluntary, self-certifying metric toolkit of best practices that designers, owners, and operators can use to guide and evaluate their progress toward high performance healing environments" (GGHC, 2006). The document has an explicit health-based focus. Before this connection, healthcare administrators often dismissed green building as having no relevance to health, viewing it as purely about saving the environment, noted Guenther. In 2002, the American Society for Healthcare Engineering (ASHE) developed the Green Healthcare Construction Guidance Statement, which articulated the need to protect health on three scales: building occupants, the surrounding local community, and the global community and resources (ASHE, 2002).

The core structure and content of *The Green Guide for Health Care* is based on existing tools, such as LEED, transferred and modified for health care. The document encourages best practices without certification and regulatory thresholds, and it bridges design and construction with operational considerations. *The Green Guide for Health Care* is a web-based, downloadable, free, open-source tool. In January 2006, the website registrants numbered approximately 6,800, with diverse geographic distribution; about 10 percent of the registrants are from outside the United States. With more than 60 projects actively based in the pilot program, a critical mass of healthcare projects are engaging in this work, noted Guenther.

The healthcare industry is beginning to recognize the relationship among medical services, the environment, and diseases. The first step was taken in 1998 when the American Hospital Association signed a voluntary memorandum of understanding with the U.S. Environmental Protection Agency pledging reductions in solid waste; avoidance of persistent, bioaccumulative, and toxic compounds; and virtual elimination of the use of mercury by 2005 (AHA, 1998).

Guenther concluded by noting the responsibility of the healthcare industry to build and operate sustainable buildings, if for no other reason than for the health of the building's occupants. The healthcare industry has the opportunity to be a model for health-based, sustainable approaches to construction, food service, active living, and waste management. It should lead creative thinking and draw inspiration from such visionaries as the early environmentalist David L. Lawrence. As mayor of Pittsburgh, Lawrence implemented a dedicated urban renewal plan connecting sustainable buildings to health and pollution prevention. Kaiser Permanente is another visionary in health care; the company's leadership in the industry is further discussed in Chapter 6.

BUILDING GREEN AND INTEGRATING NATURE: RIKSHOSPITALET UNIVERSITY HOSPITAL IN OSLO

Building green requires a big picture approach. At its core, planning and building green hospitals require that little harm be done to the macro- and microenvironments, noted Knut Bergsland of SINTEF Health Research. This approach should be taken throughout the life cycle of the building and should include all support systems during its useful life. Healthcare facilities need to find the most important indicators for green hospital building. To establish a culture for green building, commitment must come from the top down, noted Bergsland. Without this leadership, it will be difficult to establish environmental values in hospitals.

The crucial, priority elements in hospital buildings are those that are most beneficial to health and require the least effort and use of resources. For example, access to nature is important to well-being. It is a deep-rooted human need that may even transcend cultural barriers, noted Bergsland. The importance of nature as a stress-reducing factor is long established (Ulrich, 1991). Thus, planning for maximum daylight and integrating nature into hospital design by as many means as possible are the right things to do in both the patient and work environments, asserted Bergsland.

> Access to nature is important to well-being. It is a deep-rooted human need that may even transcend cultural barriers.
>
> —Knut Bergsland

Success Story of Green Hospital Building in Norway

Norway is a small country on the outskirts of Europe with 4.6 million inhabitants (approximately the population of Colorado) and a population density similar to that of Maine. Norway's per capita gross domestic product is similar to that of the United States at approximately $48,000 (CIA, 2007), and 10 percent of it is spent on health care (Johnsen, 2006). According to Bergsland, the healthcare system in Norway is driven by the same forces as in most other Western countries—demographic change, new information and technologies, and demands for efficiency. Health care in Norway is delivered through a national system based on equal access to and distribution of services as the main principle. The Norwegian healthcare system is 90 percent public and tax based. Hospitals are owned by the state and run as trusts, inpatients do not pay for their stay, and physicians are employed by the hospital.

Rikshospitalet University hospital in Oslo has 1.5 million square feet of floor space and is located next to woodlands with views of the city and the Oslo fjord, noted Bergsland. It is a tertiary teaching hospital providing world-class medical services, such as transplant surgery. The hospital covers all clinical specialties except for geriatrics and psychiatry; it has 585 beds, excluding intensive-care beds. The hospital has 4,000 full-time-equivalent staff members, 35,000 inpatients, 20,000 day patients, and 160,000 outpatients per year. Between 2000 and 2004, inpatient activity in the Rikshospitalet increased by 22 percent, and outpatient activity increased by 66 percent (Bergsland, 2005).

The Rikshospitalet site—on cultivated land next to the existing medical faculty—was selected by the Norwegian Parliament over protests from environmental activists. The Rikshospitalet counters the notion that hospitals should be built on a flat site, noted Bergsland. Because the site is sloping and saucer shaped, it effectively hides the substantial building structure; big volumes can be hidden in the bottom of the saucer and make the hospital appear as a 3- to 4-story building, while it actually is 6 to 7 stories high. The hospital sits on 87 acres of land, the footprint for the entire construction is 430,560 square feet, and the main building is 322,920 square feet, noted Bergsland.

The architects' vision for the physical environment of the hospital was a village-like horizontal layout with daylight in all spaces. The architects sought to reduce anxiety and build dignity for both patients and personnel, using natural materials whenever possible (Figure 2-2). The hospital planning process started in 1990 and involved 800 people, more than 15 percent of the total workforce, including medical staff. Because of the village concept, the hospital is seen as a town rather than a building: it has a main street, a square, and a landmark tower.

The main street facilitates way finding because intersections are unique, not identical, as they tend to be in most modern hospitals. The Rikshospitalet's curving main street is a device for patients to draw a mental picture of the route to their destination. The curvature hides the length of the corridor—280 meters long—and the traffic hierarchy minimizes the need for signage. The art and nature at the intersections help people remember their location and aid recognition. Recognition facilitates the trip, and the shortest distance to one's destination is not a straight line, but the most beautiful route.

The main street is also a place of positive distractions; for example, concerts are held on the street at least once a month. Art is an integrated part of the building; 0.9 percent of the total building budget was earmarked for art in the hospital. Art may have similar effects on stress reduction as nature, noted Bergsland. The main plaza of the hospital faces south, creating sunny spots along the perimeter. Norway is a cold country with a low sun, and the extra light is welcome.

Integrating nature was part of the hospital design; there is a walking trail next to the site, with a creek running between the trail and the hospital (Figure 2-3). Some patients use the woods next to the hospital as part of their therapy, said Bergsland. Natural stone was used in the main street, the floors, and street furniture; wood was used in benches, chairs, reception desks, and the cafeteria. Great care was taken to preserve existing trees. Views to the outside from the main street open up to green spaces and courtyards, and there are places for contemplation. This feature also illustrates that Rikshospitalet is a friendly, nonfrightening building. It is very distinct from the big clinical machines that are commonly seen, noted Bergsland.

The main entrance of the central plaza is a tower that creates the first impression of the hospital. According to Bergsland, the first impression of a building plays a disproportionate role in the conception of what comes after, such as com-

FIGURE 2-2 The village structure of the main street at Rikshospitalet aids recognition and facilitates way finding. Integrated art serves as a stress reducer. Natural daylight creates space efficiency and evokes a positive response from patients and staff.
SOURCE: Rikshospitalet Information Department, unpublished (2005). Reprinted with permission.

FIGURE 2-3 This hospital's architects focused on designing a humanizing environment, not minimizing the footprint. The main street is the backbone of the hospital and encourages informal meeting between staff. Walkways connect patient units with treatment facilities.
SOURCE: Rikshospitalet Information Department, unpublished (2005). Reprinted with permission.

munication with a doctor. The staff cafeteria is located next to the main entrance, which facilitates informal meetings among staff members. Walkways on three levels connect the patient units with treatment facilities across the main street. Inpatients are taken to treatment across the street in their beds.

The hospital is accessible by public transportation. Oslo city authorities extended an existing rail track when the hospital was planned, and today connections to central Oslo run every eight minutes. For those who drive to the hospital, the parking structure is a 4-minute walk to the main entrance. Bicycle parking is located outside the hospital, under the main plaza.

Energy Use in the Rikshospitalet

In terms of energy use, the Rikshospitalet is not an exemplar project, noted Bergsland. It uses more energy per square meter than most other Norwegian hospitals. Increased clinical demand has resulted in a need for extra capacity and ventilation, and some system limits have been passed. However, the glassed roof brings Norwegian winter light into the main street of the hospital, and the extra energy that is required to keep the street at 17°C in the winter is more than outweighed by the positive effect on staff morale, said Bergsland. The hospital tries to be environmentally conscious about its energy use, and, despite a 20-percent increase in clinical services in 3 years, it reduced its energy use by 10 percent (Bergsland, 2005).

The positive feedback from the people who are using the building more or less corroborates the concept that the architects suggested, said Bergsland. A preliminary study on the effects of hospital design on patient attitudes, activity patterns, productivity, and staff morale at the Rikshospitalet was performed in 2004. The results showed that people liked the building because it was interesting and nonfrightening, and they thought the main street was perfect for interaction (Bergsland, 2005). Among other positive factors cited was daylight in working and patient spaces and good functional proximities between related departments. Also, the art made staff feel proud of their environment.

Patients ranked the Rikshospitalet highly. Furthermore, productivity measures increased, and absenteeism and turnover rates decreased. The average sick leave in Norwegian hospitals is approximately 8 percent. After moving to the new building, the Rikshospitalet personnel's sick leave rate declined from 8 to 6 percent (Bergsland, 2005).

The building concept may have played a role in achieving patient and staff satisfaction, said Bergsland, but is difficult to determine the role of design on activity, productivity, or medical outcomes. Such factors as the Hawthorne effect,[†] moving into new premises, organizational changes, and staffing levels may influence outcomes to a degree that is difficult to establish.

[†]An increase in worker productivity produced by the psychological stimulus of being singled out and made to feel important.

3

Economics, Ethics, and Employment

During the workshop, speakers and participants considered the economic and ethical driving forces for a green building agenda, and the concept of expressing organizational ethics in building design. This chapter summarizes presentations from three speakers: Gregory Kats, John Poretto, and George Bandy. These speakers described current research and provided insight based on their personal observations and experience. Future goals and research needs in this area are discussed.

GREEN BUILDING: ECONOMICS

Businesses embrace the idea of green buildings because they believe they are ethical, productive, and healthy, noted Gregory Kats of Capital E. The primary driving forces are quality of life and health. He remarked that many American corporate headquarters (e.g., Goldman Sachs, Reuters, New York Times, Bank of America) are building green facilities, often more proactively than other sectors, such as government, residential housing, and academia.

Building a green facility involves following guidelines, such as Leadership in Energy and Environmental Design (LEED), the nationally recognized green building rating system developed by the U.S. Green Building Council. LEED is not an exact science, but rather a consensus-based approach to defining practical criteria for green building. LEED is the current best practice standard for the building sector, although it is somewhat subjective in its criteria. For example, energy professionals believe that energy is underweighted, and those in the water field believe the same is true for water. Because of this, it is important to quantify the benefits objectively, asserted Kats.

A recent study aggregated the cost and financial benefits of 33 green buildings (Kats and Capital E, 2003). Although commissioned for California, it had a national focus. In general, the study reported that the initial construction cost

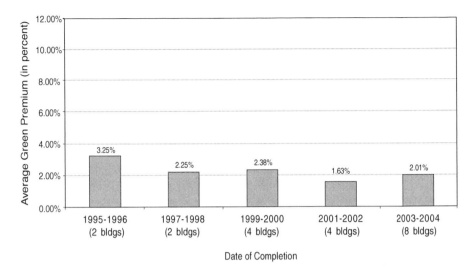

FIGURE 3-1 The average cost premium associated with building green for U.S. buildings certified at the LEED Silver level has generally decreased over time. This is primarily due to familiarity with the process and the requirements of being LEED certified.
SOURCE: Modified from Kats and Capital E (2003).

was higher for green buildings—approximately 2 percent, or \$3–5 per square foot—than nongreen buildings, noted Kats. However, he suggested that some of the cost premium can be attributed to the novelty of building green and the fact that many builders are on a learning curve. Construction costs decline as more green buildings are constructed and familiarity with green design increases, and when green building principles are incorporated early in the design process. For example, the average cost premium for a green building certified at the LEED Silver level has generally decreased over the past 10 years (Figure 3-1). An important message for institutions that engage in their first green building is that their next green building is likely to cost less, said Kats.

Health and Productivity Are the Drivers for the Benefits of Green Facilities

When quantifying benefits, Kats's group focused on areas with the largest potential gain. Rent or amortized ownership account for approximately 5 percent of operating costs, and the direct and indirect costs of employees constitute 80–90 percent. Thus most studies look to this area of productivity and health for the largest impacts.

In a financial benefits summary of green buildings (Table 3-1), the study found that energy benefits saved \$5.80 per square foot, while operating costs saved approximately \$8.50 per square foot. For productivity and health, four

TABLE 3-1 Financial Benefits of Green Buildings—Summary of Findings (per square foot)

Category	20-Year Net Present Value
Energy savings	$5.80
Emissions savings	$1.20
Water savings	$0.50
Operations and maintenance savings	$8.50
Productivity and health value	$36.90–55.30
Subtotal	$52.90–71.30
Average extra cost of building green	(–3.00 to –$5.00)
Total 20-year net benefit	$50–65

SOURCE: Kats (2003).

drivers were measured: lighting control, ventilation control, temperature control, and the amount of daylighting. These four drivers are a subset of a larger number of factors that affect productivity and health. For buildings certified at the LEED Certified and Silver levels, productivity and health increased about 1 percent. For buildings certified at the Gold and Platinum levels, the increase was about 1.5 percent or 7 minutes of employee time saved per day, noted Kats. The largest cost for a public or a private entity is the cost of its employees; these numbers translate into a net saving of $34–55 per square foot, depending on the level of certification. He concluded that an initial cost premium of $3–5 per square foot results in a net return of $50–65 per square foot net over a 20-year period at a 7 percent discount rate (Kats, 2003). There were additional significant financial benefits that the study was not able to quantify, including insurance, employment, equity, security, and brand appreciation.

Green Schools

Another recent study that gathered objective information on green buildings examined the costs and benefits of 30 green schools (Kats, 2006). Although the data came from a national sample of schools, the energy costs and teacher earnings were calculated on the basis of Massachusetts-specific costs. Both the conventional and the green school buildings averaged approximately 125,000 square feet for 900 students. The average cost premium for the 30 green schools in 10 states nationally was 1.65 percent, which translates into a cost premium of $3–4 per square foot. The study results showed an energy savings of 33 percent and a water savings of 32 percent (Kats, 2006). There were also substantial academic gains. In one example, two conventional schools were combined into a newly constructed green school in North Carolina. For the three years prior to the move, only 60 percent of the students achieved state-level standards on mathematics and reading; after the move, the proportion rose to 80 percent. The only changes were the introduction of good daylighting, ventilation control, temperature controls, and control of pollutants; the students, teachers, and parents remained constant (Kats, 2006). A reduction of approximately 15 percent in colds and flu was also observed. Overall, the study found a benefit-to-cost ratio of 20 to 1 in green schools.

Determining the impact of green schools on the health and performance of students is an important area, but the research has lagged. The Kats report was among the first to analyze the benefits in this area. Subsequently, the National Research Council (NRC) published a report entitled *Green Schools: Attributes for Health and Learning.* The charge to the NRC committee was to "review, assess, and synthesize the results of available studies on green schools and determine the theoretical and methodological basis for the effects of green schools on student learning and teachers' productivity" and to look at possible impacts of green schools on student and teacher health (NRC, 2006).

The committee found that a number of factors made the task more complex than might first be evident. These included the lack of a clear definition of what constitutes a green school; the difficulty of measuring educational and productivity outcomes; the variability and quality of the research literature; and confounding factors that make it difficult to isolate the effects of building design, operations, and maintenance. In addition, most inferences about the impact of the built environment on health and performance are based on studies of adult populations. Committee members noted that extrapolating from these studies to younger populations is difficult. In its review, the committee "did not identify any well-designed, evidence-based studies concerning the *overall* effects of green schools on human health, learning, or productivity," noting that this is understandable because the concept of green schools is relatively new and evidence-based studies require significant resources. The committee did find sufficient scientific evidence to establish an association between some aspects of building design and human outcomes, including acoustics and learning, excess moisture and health, and indoor air quality and health. Additional findings and recommendations on the state of research on the building envelope, indoor air quality, lighting, acoustics, ventilation, and the transmission of infectious diseases are included in the report.

Health and Productivity Gains

A review of the literature by the Carnegie Mellon University Center for Building Performance found a range of productivity gains related to both improved temperature control (Figure 3-2) and high-performance lighting systems. Kats observed that one of the characteristics of high-performance green buildings is that they typically have more sophisticated energy management systems and better integrated lighting strategies. He noted that the Carnegie Mellon group also

22

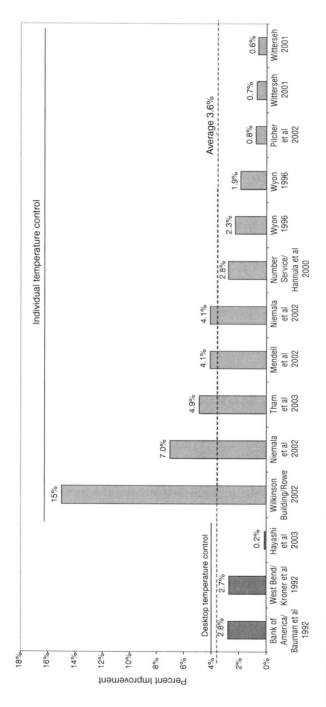

FIGURE 3-2 The average productivity gains from improved temperature control across a number of peer-reviewed studies was about 3.6 percent. The increase ranged from 0.6 to 15.0 percent.
SOURCE: Carnegie Mellon University Center for Building Performance (2005).

found health benefits from improved indoor air quality. Improvement in indoor air quality was linked with an average reduction of 41.5 percent in symptoms.

A study by the real estate firm of Cushman and Wakefield (RICS, 2005) surveyed 12 owners of public and private green buildings to determine what they thought were the most significant effects attributed to building design. Health and productivity benefits of working in a green building outranked the benefits of decreased energy consumption and operating costs. Kats views this as another indicator that the real estate community recognizes health and productivity as driving forces, although they are harder to measure than reduced energy and water consumption.

In another survey, Turner Construction (one of the largest construction firms in the country) asked 719 executives to rate the benefits of green buildings compared with nongreen buildings (Turner, 2006). Of these executives, 60 percent worked at organizations currently involved with green buildings. The survey covered a number of categories, including building value, worker productivity, return on investment, rents, occupancy rates, and retail sales. Overall, the executives rated the benefits of green buildings very highly, and those with more direct experience with such buildings were more positive in their responses. For example, of executives involved with green buildings, 91 percent believed that these buildings improved the health and well-being of occupants, compared with 78 percent of executives not involved with green buildings. Similarly, 65 percent of executives involved with six or more green buildings said the residents or occupants enjoy "much greater health and well-being," compared with 49 percent of executives involved with three to five green buildings and 39 percent of executives involved with only one or two green buildings.

Kats pointed out that green buildings are a fundamental part of addressing global warming because of their energy efficiency. There are opportunities to invest in energy-efficient technologies, such as energy-recovery ventilators and ground source heat pumps. Overall, Kats emphasized that building green is a good news story. The more experience that various sectors have with green buildings, the more costs come down. He suggested that people will increasingly recognize the benefits of building green as they continue to gain experience with these buildings.

ETHICS OF GREEN BUILDINGS

John Poretto of Sustainable Business Solutions explained that the Hill-Burton Act triggered a wave of hospital construction beginning in the 1950s. Today, many buildings from that era are becoming obsolete and are scheduled for replacement. Poretto questioned why we are building facilities that only last 40 to 50 years and that lack flexibility and adaptability. He suggested that it might be time to look for a new paradigm.

Poretto drew on his experience as the executive vice president and chief operating officer for the University of Texas Health Science Center at Houston to illustrate his points. During this time, the university owned a 4- to 5-year-old building of approximately 850,000 square feet. It was the second most expensive building to operate in the state of Texas, and within four years of its construction, the Texas State Department of Health declared it a sick building. Recognizing that the university had one of the state's largest, most costly, and most dysfunctional buildings caused him to look at green buildings in a more pragmatic, ethical, and holistic manner.

The University of Texas Health Science Center embarked on constructing a state-of-the-art green building for its nursing students with the goal of teaching them early in their careers about the benefits of building green, as well as instilling an appreciation for the green healthcare facilities in which they are likely to work during their careers. The building was financed primarily with fees assessed to students. The state legislature and the philanthropic community contributed, although Poretto noted that it was primarily the students' commitment that made this new approach possible.

Architectural Style Denotes Commitment

Poretto drew on the writings of Hippocrates, in particular an essay entitled, "The Physician," which explores the basic ethical standards and actions by which a doctor should live. In this essay, Hippocrates maintained that "the dignity of a physician requires that he should look healthy, and as plump as nature intended him to be; for the common crowd, consider those who are not of this excellent bodily condition to be unable to take care of others." Poretto questioned why an essay that later espouses ethical ideals, such as honor and trust, opens with a focus on the doctor's physical appearance. He interpreted this idea to mean that the first thing a patient encounters during a medical consultation is the physician's physical appearance and bearing, which provides important insight about the physician's health, health behavior, and trustworthiness. In an analogous way, when people visit an organization, the first attribute they notice is the building that houses it. For those who learn and work in clinical and health education institutions, the design and architecture send crucial daily messages about the institution's identity and values and about attitudes toward the health and productivity of the building's occupants.

Poretto asserted that most healthcare buildings do not convey a message of humanistic concern for the welfare of others. He called for a more "principle-centered" and responsible operational model, one that reflects the highest values of both donors to, and leaders of, health institutions. This model requires attention to a healthcare facility's purpose, use, design, construction, and maintenance, and it needs to extend holistically from staff within the institution outward to the community as a whole.

Poretto focused attention on the tax-exempt status of many healthcare institutions, a status that reflects an institutional obligation to serve the good of the community as a whole. A tax-exempt institution's mission has a higher meaning than that of profit. If, indeed, it is the goal of healthcare institutions to serve human needs and well-being, then an understanding of how best to meet expectations is necessary, asserted Poretto. The leaders and the people selected to carry out this lofty mission should be competent and compassionate and possess an ethical purpose that grounds their leadership. High objectives and relevant standards must be set for selecting people who lead and work in the facilities, noted Poretto. He suggested that the organization's actions cannot be based on values different from those that are taught to students or applied to patients. These expectations support the notion of green healthcare facilities.

Long View of Ethical Principles

Building green facilities does not stop with the organization; it also requires a holistic view of its operations, noted Poretto. He pointed out that organizations should work with entities that share and complement their ethics, visions, and principles. As tax-exempt organizations, these entities must be held accountable for creating and maintaining principle-centered operations and for behaving in ways that demonstrate wise expenditures. Poretto asserted that decisions guided by short-term considerations lead to preventable problems and costs, which are usually far greater than preventive measures.

> Decisions guided by short-term considerations lead to preventable problems and costs, which are usually far greater than their preventive measures.
>
> —*John Poretto*

What it means to be green or sustainable and how that is consistent with the expectations of a tax-exempt status start with a multigenerational viewpoint. The principal difference between ethics and policy that focus on the individual and ethics and policy that focus on the institution is the latter's multigenerational potential and perspective. He noted that this requires constructing buildings that have the capability to span generations. He emphasized the need for early discussion of several issues, including how to gain the greatest efficiency from the natural surroundings, how to take advantage of daylighting and natural breezes, protection from inclement weather, and so on. Poretto asserted that green buildings accomplish this because they use natural materials, are flexible, and are multigenerational in their approach. Higher education and healthcare associations have outstanding yet currently underrealized opportunities for bringing about volume pricing breaks and providing support for smaller companies to offer competitively priced green goods and services. He concluded by saying that buildings of healthcare institutions should be seen and evaluated not only as a functional elements of work and symbols of social or artistic standing, but also as an ethical statement by the institutions. This statement should demonstrate concern and responsibility for the well-being of the building's users and the community.

INCREASING WORKPLACE PRODUCTIVITY

George Bandy of Interface Research stated that business may have framed the phrase "worker productivity" too narrowly. The sciences connected to productivity are ergonomics, cognitive psychology, social psychology, cultural psychology, ecology, biology, economics, leadership, and management. Ergonomics and cognitive psychology relate to the *I*. Social psychology and cultural psychology relate to the *we*, and economics, leadership, and management relate to the *they*. He asserted that the challenge is to convert the *they* to the *we*. The *I* and the *we* are the users and the occupiers of the building. He suggested that the industry sector needs, through creativity and learning, to identify the population that they are servicing. Interaction and dialogue are essential to this effort. The challenge is to determine what should be researched, for whom and why, and what should be taught now so that in the future professionals are well prepared. He suggested that careful studies of human behavior are needed, the design of which should focus on goals, health, and productivity.

The commitment to sustainable development is an ethical decision requiring a conscious choice to provide for the needs of present and future generations, noted Bandy. The challenge is to develop a strategy that creates healthcare professionals who recognize and choose to create ecologically sound technologies for healthcare institutions and the communities they serve. He suggested that a strategy is needed that inspires more sustainable behavior, not just good intentions. An individual's behavior is influenced by his or her knowledge of facts and the values and norms of his or her environment. The primary mission is to cultivate knowledge in students and practitioners so they can care for the health needs of individuals. A central focus of recent research is the influence of the environment on the health and wellness of individuals and communities, observed Bandy.

> A strategy is needed that inspires more sustainable behavior, not just good intentions.
>
> —*George Bandy*

Environmental health relates to more than the natural environments of air, water, and land; it also encompasses the built environment, noted Bandy. The environment in which one trains employees, negotiates contracts, and performs surgery reveals much about society's collective human principles. He asserted that a company cannot profess to be genuine in its aims for worker productivity, health, and well-being if it conducts business and provides services for clients and facilities that are unhealthy and economically wasteful. Bandy asserted that workplace environment matters, noting that there are five key factors in terms of the working environment: personal space, climate control, daylight, office design, and quiet facilities. He concluded that the evidence suggests a need for sustainable development, especially for healthcare facilities, and further suggested that developed countries can provide leadership in sustainability and serve as a model for developing countries.

4

The Health Aspects of Green Buildings

The previous chapter discussed green healthcare institutions in the framework of economics, ethics, and employment. This chapter explores the relationship between green healthcare institutions and human health. It includes information from four presentations by professionals with a broad range of expertise: Todd Schettler, Anthony Bernheim, Judith Heerwagen, and Derek Parker.

Buildings are complex and dynamic systems producing a heterogeneous indoor environment consisting of many microenvironments. Many factors, including temperature, humidity, light, noise, chemical pollutants, odors, personal health, job or activity requirements, and psychosocial factors, interact to influence the comfort and health of building occupants, said Ted Schettler of the Science and Environmental Health Network. The interactions among these factors and the dynamic heterogeneity of microenvironments within buildings, including temperature and humidity gradients, make the indoor environment a complex area to study, noted Schettler. However, many indoor air quality studies are not designed to address these complexities, thus contributing to the conflicting information that is often found in the scientific literature. To better understand the health impacts of indoor environments, new statistical techniques, such as principal component analysis, which consider multiple variables simultaneously, are necessary, said Schettler.

HEALTH IN BUILDINGS: INDOOR AIR QUALITY

According to studies by the Environmental Protection Agency (EPA), Americans spend an average of 90 percent of their time in an indoor environment in which low air circulation can concentrate pollutants to two to five times higher than in outdoor air (EPA, 2006). Chemicals that define indoor air quality can also affect health. Outside air, which comes through natural ventilation as well as mechanical systems, contributes to the quality of the indoor air. The tightness of the building determines the circulation of air (and air contaminants) in and out of the building. If the ventilation system is not efficient enough at dissipating the pollutants that are brought into the building, they will stay there longer, thus affecting health, said Anthony Bernheim, principal of green design at the architectural firm of SMWM. According to Bernheim, several major factors affect indoor air quality: the quality of the outside air, the location of outside air intakes, construction materials, furnishings, equipment, filtration and ventilation efficiency, occupants, and maintenance. The following types of chemical compounds are found in indoor air:

• Volatile organic compounds, which may be emitted from building materials and fabrics, new furniture, cleaning materials, vinyl wall coverings, and office equipment
• Microbial volatile organic compounds, such as mold and mildew
• Semivolatile organic compounds, which come from fire retardants and pesticides
• Inorganic gases, such as ozone, carbon monoxide, and nitrogen dioxide
• Particulate matter from burning fuels in cars and from burning combustion products

In the early 1990s, a Scandinavian researcher, Lars Molhave, introduced the term *total volatile organic compounds* to reflect the burden of volatile organic compounds in indoor air. This term has received considerable acceptance among indoor air quality experts, noted Bernheim (Hudnell et al., 1992).

Building-Related Health Effects

Building-related illnesses include symptoms associated with sick building syndrome as well as specific building-related illnesses, such as Legionnaire's disease. Symptoms associated with sick building syndrome include headache, nausea, nasal and chest congestion, wheezing, eye problems, sore throat, hoarseness, fatigue, chills and fever, muscle pain, and neurological symptoms, such as difficulty remembering or concentrating, dizziness, and dry skin. These symptoms do not necessarily keep people away from work, but they are often a source of complaints and undoubtedly contribute to lost productivity and dissatisfaction with the work environment, said Bernheim. The symptoms are not easily attributable to any particular chemical in the building and generally subside when people leave the building.

Sharp distinctions between health and comfort are not always readily apparent and may not be appropriate. Attempts to draw distinctions contribute to contradictory and inconsistent research findings, noted Schettler. Many researchers in this field think that complaints of building-related symptoms are worthy of investigation, even if a definable disease cannot be identified, said Schettler.

Sharp distinctions between health and comfort are not always readily apparent and may not be appropriate.

—*Ted Schettler*

Some building-related diseases, such as Legionnaire's disease, can ultimately be traced to a well-defined cause that can be addressed. More often, building occupants experience a variety of vague symptoms that may change over time, which makes their analysis difficult, noted Schettler.

As an added complexity, some people in the general population seem to be disproportionately sensitive to various environmental exposures. Multiple chemical sensitivity is a condition in which people report sensitivity or intolerance to a number of chemicals. According to Schettler, multiple chemical sensitivity is somewhat controversial because its pathophysiology, the natural history of the disease, and how to respond to it are not well understood. Nonetheless, an increasingly robust scientific database supports the importance of this phenomenon.

Design Principles in Healthy Buildings

It is important to design and build in ways that reduce the probability of mold growth, avoid moisture accumulation, consider cleaning requirements, and reflect an understanding of the influence of factors such as low-emitting materials, ventilation, humidity control, and surface temperatures on indoor air quality. It is important to understand buildings and the indoor environment as complex dynamic systems, as well as to consider the full life cycle of materials, said Schettler.

According to Bernheim, there are four principles for good indoor air quality design:

1. Source control: keep the source of pollutants out of buildings or reduce the sources when they cannot be prevented.
2. Ventilation control: provide adequate ventilation to dissipate the pollutants and get them out of the building.
3. Building commissioning: define performance specifications in advance and test the building at various stages of construction and operation in order to ensure that it performs as designed.
4. Maintenance: ensure that the building is kept clean and maintained during its operational life.

Bernheim observed that source control has been the focus of most of the indoor air quality initiatives in the past 5 years. In 1981, the American Society of Heating, Refrigerating, and Air Conditioning Engineers created guidelines for source control, which were used until the early 1990s. In 2000, the firm of SMWM, in collaboration with the state of California, created health guidelines based on chemicals of concern such as carcinogens, reproductive toxicants, chemicals with acute reference exposure levels (ARELs), and chemicals with chronic reference exposure levels (CRELs). ARELs refer to 1 to 7 hours of exposure, and CRELs refer to approximately 12 to 15 years of exposure. The California EPA created a list of CRELs that includes 80 chemicals commonly found in buildings (Lent, 2006).

These CRELs can be linked to the standard industry format for building specifications, the Construction Specifications Institute's MasterFormat™. MasterFormat™ is structured as a standardized outline form with 16 divisions. For example, Division 1 contains general administrative and procedural requirements, and Divisions 2 through 16 address technical specifications for building materials. A section on environmental protection procedures can be added to Division 1. This section, often referred as Section 01350, provides a forum for identifying environmental requirements, such as sustainable site planning, construction recycling, energy efficiency, indoor air quality, and others (CIWMB, 2007). Bernheim described a Section 01350 requirement that a building may not expose occupants to more than half of the material's CREL. Based on the quantity of material used in the project and the volume of air in the system, analysts produce a modeled concentration that is matched against the Section 01350 list, said Bernheim.

Using existing data has enabled researchers to begin to analyze more closely what is happening in a building. According to Bernheim, the Section 01350 specification has led to significant industry transformation. For example, based on Section 01350 testing, a national ceiling tile manufacturer has completely modified its ceilings to reduce formaldehyde emissions, and an international manufacturer of linoleum flooring has reduced emissions of chemicals from linoleum. As another result of Section 01350 testing, trade organizations for building materials created their own certifications, such as the Carpet and Rug Institute's Green Label Plus Carpet program for the carpet industry, and the Resilient Coverings Institute's Floor Score certification for flooring. The Collaborative for High-Performance Schools in California has used Section 01350 as a guideline. *The Green Guide for Health Care* referenced Section 01350, as did version 2.2 of the U.S. Green Building Council's Leadership in Energy and Environmental Design (LEED) for new construction, said Bernheim.

While concentrating on the indoor environment, green healthcare advocates also need to understand public, occupational, and environmental health impacts beyond the building. Such issues as materials extraction, manufacturing, transport, and disposal have potential health effects for people and communities, noted Schettler.

Beyond building green, the healthcare industry has a responsibility to address its contribution to unsustainable material throughput, growth, and natural resource depletion and degradation. In their book *The Ethics of Environmentally Responsible Health Care,* Jessica Pierce and Andrew Jameton (2004) argue that health care has a particular ethical responsibility and that marginal improvements in

Beyond building green, the healthcare industry has a responsibility to address its contribution to unsustainable material throughput, growth, and natural resource depletion and degradation

—Ted Schettler

building materials' policies are insufficient. A fundamental reexamination of the scope of clinical services is also required if health care is to do its part to restore and maintain resources and ecosystems on which life depends. This creates concerns about healthcare rationing; according to Pierce and Jameton, rationing should be thought of as sustainable optimal care rather than less than optimal care in order for the healthcare industry to meet its ecological responsibilities.

Schettler added that it is important to think about how the greening of medical care might be introduced in medical education. Medical students should understand the links between individual health, community health, and ecological health in a way that helps to develop an integrated ecological consciousness.

SUSTAINABILITY, HEALTH CARE, AND PATIENT WELLNESS

In her presentation, Judith Heerwagen of J.H. Heerwagen and Associates drew heavily from Australian biologist Steven Boyden's theory of human ecology. Boyden (1971) defines biological determinants of optimal health as "those various conditions which tend to promote or permit optimal physiological, mental, and social performance in an animal in its 'natural' or evolutionary environment." Boyden argues that environments need to fully satisfy both survival needs and well-being needs, which are different. Survival needs have to do with clean air and water—people are very likely to get sick without these assets—while well-being needs have to do with psychosocial adjustment, stress reduction, and quality of life (Boyden, 1971). There are several evolved well-being needs and experiences. Heerwagen described scientific evidence that social support is connected to being healthy. Neuroscientists are learning that creativity has been a survival function in the evolution of the human brain, suggesting that opportunities for creative activity are also very important. Variety in daily experience also enhances well-being, as does behavioral control (the ability to react and adjust one's behavior in response to different environments). People need an interesting and aesthetically pleasing environment, sensory stimulation similar to that found in the natural environment, and connection to the natural world. These needs may vary slightly across different age groups and different healthcare problems; however, they are relevant to most healthcare environments and contexts, reflecting people's need to be healthy in a psychosocial sense.

Psychological and Social Aspects of the
Environment in Healthcare Facilities

In a qualitative survey of 50 hospital inpatients in the United Kingdom, participants identified a need for a hospital environment with personal space; a homey, welcoming atmosphere; a supportive environment; good physical design; access to external areas; and facilities for recreation and leisure (Douglas and Douglas, 2004). These results demonstrate the need for attention to the psychological and social aspects of the healthcare facility environment. Currently, hospital environments confront patients with psychosocial deprivation that creates negative health consequences, said Heerwagen. Patients in healthcare facilities experience pain, discomfort, and anxiety. Boredom is a very common complaint in hospitals, reflecting the lack of creative activity and mental stimulation. According to the survey, patients also say that isolating hospital environments lead to a loss of emotional support, social support, and behavioral control, noted Heerwagen. The television is often the only thing that patients can control in most healthcare facilities. Generally, they cannot control the thermal environment, lighting, or ventilation. There has been a great deal of research in psychology that examines increased patient control in interactions with physicians. Patients who take part in decision making and become more informed report an increased sense of control. Thus green healthcare advocates should consider whether greater control over the environment could contribute to positive medical outcomes, said Heerwagen.

According to Heerwagen, sunlight in healthcare facilities is associated with substantial reductions in medical costs. Researchers who study the benefits of sunlight found evidence that lighter and brighter rooms in hospitals contribute to stress reduction and that patients experience less pain and use less analgesic medicine (Walch et al., 2005). Studies involving patients with depression or bipolar disorder have shown that sunlight in patient rooms contributes to shorter hospital stays and reduced symptoms (Beauchemin and Hays, 1998; Benedetti et al., 2001). Further evidence suggests that people with seasonal affective disorder prefer more brightly lighted spaces and that such spaces are linked with the reduction of their symptoms (Eastman et al., 1998). A study by Beauchemin and Hays (1998) showed a reduced mortality rate among heart attack patients who were hospitalized in bright, sunny rooms.

Technology, rather then nature, is the main source of stimulation in a hospital setting. Hospitals are rather noisy places, noted Heerwagen, and the sounds in hospitals are primarily technical, because they are used to provide signals to caregivers about patient status. These noises are particularly constant in intensive-care units and in postsurgical or postanesthetic units. These noises are amplified by the hard surfaces and the lack of acoustical tiles and treatment in most hospitals. Although carpeting has a real acoustical value, nurses and maintenance workers dislike it because it is more difficult to clean.

Noise has been associated with disturbed sleep in patients (Topf, 1992);

increased stress among staff, particularly in intensive-care units (Blomkvist et al., 2005); and headache, irritability, and increased sensitivity to pain (Biley, 1994). A study by Shertzer and Keck (2001) found that patients perceived less pain when noise was reduced and replaced with music. Heerwagen stressed that healthcare facility leaders should consider the impact of noise as they engage in building design.

Theory of Positive Design

A key challenge for specialists in healthcare facility design is how to increase a sense of well-being without compromising ongoing medical care. Conflict arises when decisions that may improve the psychosocial situation interfere with caregiving. Knowing how, when, and for whom to provide psychosocial stimulation is critical, noted Heerwagen.

> A key challenge for specialists in healthcare facility design is how to increase the sense of well-being without interfering with ongoing medical care.
>
> —Judith Heerwagen

Improving the psychosocial state of patients is an important consideration in hospital design, she said. Hospitals should address the following key environmental factors: aesthetic pleasantness of the building, sunlight, noise reduction and positive sound stimulation, connection to nature, socially supportive spaces where patients can be with family, and increased behavioral control. The first principle in the theory of good design is reducing health and safety risks. Creating an atmosphere that is supportive psychologically, cognitively, emotionally, and socially is also important and should be incorporated in positive design principles, said Heerwagen. Reduction of noise (which acts as a stressor) and the use of music therapy to enhance the patients' well-being are one example. The benefits of positive design include reductions in pain, emotional anxiety, and other physiological indicators of stress (Cabrera and Lee, 2000).

Quality improvements to the hospital environment may involve costly additions to space and furnishings. However, this should be viewed as a cost-benefit trade-off that has other values, said Heerwagen. Aesthetic pleasantness reduces the "institutional" atmosphere and makes people feel that they are valued and worth investing in. Job satisfaction is much higher in places with aesthetically pleasant environments, noted Heerwagen. The links between sustainability and psychosocial outcomes are clear; however, there is still much to learn in this area, she concluded. She closed by emphasizing that "it is not how green you make it that counts, but how you make it green."

> It is not how green you make it that counts, but how you make it green.
>
> —Judith Heerwagen

DESIGN RESEARCH AND THE BUSINESS CASE FOR A BETTER 21ST CENTURY HOSPITAL

The Fable Hospital

In an attempt to find out how much a better building would cost, Derek Parker of Anshen + Allen Architects and his associates invented the imaginary Fable Hospital to measure experience using evidence-based design. The Fable Hospital was based on the Center for Health Design's Pebble Project, a research program that was initiated with San Diego Children's Hospital and Health Center in 2000. Located on a limited urban site, the Fable Hospital provides a comprehensive range of inpatient and ambulatory services, including medical/surgical, obstetrics, pediatrics, oncology, cardiac, and emergency medicine. It was built to replace a 300-bed regional medical center at a cost of $240 million. The hospital values quality, safety, patients, families, staff, cost, value, and community responsibility. In fact, the hospital collaborates with the Institute for Healthcare Improvement and has an unusual culture; it is obsessed with quality and safety, driven by values, patient focused, family friendly, a good corporate citizen, determined to be ecosensitive, willing to benchmark, and committed to being held accountable.

The Fable Hospital client is data-driven and engages in participatory planning as it designs its new facility. A wide range of stakeholders was engaged, including not only hospital management and staff, but also architects, engineers, interior designers, contractors, and landscape architects. Participants considered evidence-based ways to improve safety and performance, improve patient satisfaction, and save money. The resulting design features readily available hand-washing stations, improved air filtration systems, better separation of "clean" and "dirty" areas on patient floors, transportation modalities that separate patients from potentially infectious materials and wastes, standardization and consistency of layout, and glare-free lighting. Other innovations include oversized, windowed, single rooms with dedicated space for patient, family, and staff activities and sufficient capacity for robots and in-room surgery. Patient rooms and work spaces have plenty of daylight. Variable acuity rooms are standardized in shape, size, and headwall; this reduces errors and eliminates the need to move patients as their condition improves.

There are decentralized, barrier-free nursing stations, computerized order entry using a bar code system and handheld computers, plentiful hand-washing facilities, and high-efficiency particulate absorbing filters. The Fable Hospital also has healing art, music, gardens, consultation spaces, a patient education center, and staff support facilities, noted Parker.

As with most hospitals, the Fable Hospital consumes large amounts of power and confronts such pathogens as *Staphylococcus aureus, Candida albicans,* and *Enterococcus faecalis.* It produces large amounts of solid, medical, contaminated, and hazardous waste. This waste has to be stored and transported or recycled using fuel cells. Currently, fuel cells are not economical; however, this could change if the money spent on disposing of waste was spent instead on treating waste as energy in a different form. Waste can become heat and power, and can produce commercially viable and ecologically sound products (mostly carborundums and additives to concrete and asphalt) that are never burned in their life cycle, stated Parker.

Detailed Cost and Savings Estimates for the Fable Hospital

Fable Hospital has private patient rooms that are 100 square feet larger than typical hospital rooms. At a cost of $185 a square foot, the larger rooms increase construction costs by $4.7 million. Overall, the construction cost premium for Fable Hospital is $12 million, or 5 percent of the construction budget, said Parker. A hospital chief executive officer would require evidence of benefits before approving such additional expenditures.

Reduction of patient falls is one of the benefits found by the Pebble Project. When patients (especially elderly patients) fall, they risk fractures and complications, such as pneumonia, that result in longer hospital stays. According to Parker, the average cost of an unlitigated fall in the United States is $10,000. A Pebble Project study found that an 80 percent reduction in patient falls can be achieved by installing double doors in bathrooms and moving telephone cords and nurse-call cords out of the way (Hendrich et al., 1995).

Based on the Pebble Project results, Parker suggests that better design may result not only in fewer patient falls, but also in fewer patient transfers, fewer nosocomial infections, reduced nurse turnover, and reduced drug costs. Based on these savings, the initial investment of $4.7 million would be recovered in a few years (Table 4-1). Parker further asserted that increasing market share and philanthropy would add to the hospital's revenues, thus justifying the construction premium (Table 4-2). Cost avoidance savings alone, if invested at 3 percent for 30 years, would pay the capital costs of the hospital many times over.

According to Leonard Berry's book *Discovering the Soul of Service*, leading

TABLE 4-1 One-Year Savings on the Fable Hospital

Fewer patient falls	$2,452,800 (−80%)
Fewer patient transfers	$3,893,200 (−80%)
Fewer nosocomial infections	$80,640 (−4/m)
Reduced nurse turnover	$164,000 (14–10%)
Reduced drug cost	$1,216,666 (−5%)
Total cost savings	$7,807,306

SOURCE: Berry et al. (2004). Reprinted with permission from The Center for Health Design.

TABLE 4-2 One-Year Revenue Gains of the Fable Hospital

Market share increase		$2,168,100
Increased philanthropy		$1,500,000
	Total revenue gain	$3,668,100
	+Total 1-year savings	$7,807,306
	Total	$11,475,406

SOURCE: Berry et al. (2004). Reprinted with permission from The Center for Health Design.

service organizations have nine drivers of success: strategic focus, executional excellence, control of destiny, trust-based relationships, investment in employee success, acting small, brand cultivation, generosity, and value-driven leadership (Berry, 1999). Although the Fable Hospital has not been built yet, it will be, said Parker. They are close to achieving that goal. Parker closed with the hope that this workshop has helped contribute evidence to make a powerful business case for quality, safety, and sustainability in American health care.

5

The Process of Change

The process of change from traditional building to incorporating green building practices can be viewed as a moving target that evolves with the institution
and the current state of knowledge. To understand the process, one should identify
who or what the institution is, how the process of change began, why the process
started, and what the challenges are, noted Bahar Armaghani of the University
of Florida. During the workshop, these concepts were discussed in terms of the
commitment to change building practices at the University of Florida and at
Emory University.

UNIVERSITY OF FLORIDA

The University of Florida has an ecological footprint of approximately 2,000
acres, more than 300 held in conservation. The campus has approximately 1,900
buildings with a combined total of 20 million square feet. Approximately 49,000
students attend the university, which employs approximately 800 staff. The University of Florida is essentially a city within a city, with ownership of the utilities
distribution system, wastewater treatment plant, and so on, noted Armaghani.
The university pays approximately $2.7 million for electricity and $85,000 for
water each year. The facilities generate about 18,000 tons of waste annually for
basic operations. During home football games, an additional 9 tons of waste are
generated at the stadium, and 7 tons from the tailgating events across the campus,
explained Armaghani. As a small city, the University of Florida believes that it
should be responsible for its use of resources.

What Are Sustainable Buildings?

According to Armaghani, sustainable buildings meet high standards in siting,
orientation, design, construction, and energy efficiency—and all of these elements

are measurable. Sustainable buildings are better for the environment
and their occupants than nonsustain-
able buildings. This can be illustrated
by comparing similar buildings after
construction. For example, of two
buildings on the University of Florida
campus of similar size and functionality, the green building is approxi-

Sustainable buildings meet high standards in siting, orientation, design, construction, and energy efficiency—and all of these elements are measurable.

—Bahar Armaghani

mately 37 percent more efficient (Figure 5-1) than a building that employs
standard construction principles. Premium costs incurred during construction are
recovered with the savings accrued by operating a sustainable building.

Why Sustainable Buildings? Health, Economics, and Environment

From a visit to the Department of Energy and the U.S. Green Building Council websites, one can appreciate the studies that illustrate the negative impacts
of buildings on the environment, including the amount of energy consumed.
The university knows there are environmental, economic, and productivity benefits to erecting sustainable buildings. During the workshop, there were discussions about improving the quality of air, water, and the environment in general,
observed Armaghani. She noted that people spend an average of 80–90 percent

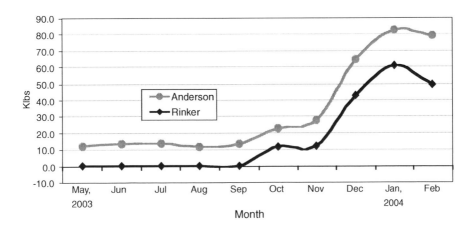

FIGURE 5-1 In comparison to traditional building standards (Anderson building), a
LEED-certified building (Rinker) was 37 percent more energy efficient. These buildings
are of similar size and functionality. Because buildings have a life span of approximately
100 years, the saving can be greater than any premium costs during the building phase.
SOURCE: University of Florida, Facilities Planning and Construction (2005, unpublished).

of their time inside buildings, and this provides motivation for institutions to minimize environmental and health impacts. Adopting the Leadership in Energy and Environmental Design (LEED) initiative was an important strategy for the university. Among other benefits, building sustainably gave the university a positive environmental image, elevating it to a position of leadership in the field, said Armaghani.

Sustainable buildings have lower utility bills, enhance assets, and increase value. She explained that the first building certified at the LEED Gold level cost the university about 7 percent more than a traditional building, but the expense was recovered in approximately 10 years. With an average life span of approximately 100 years, the university considers the initial capital investment in green buildings to be justified, noted Armaghani.

Armaghani observed that building sustainably provides benefits to public health, including improved air quality, minimized strain on infrastructure, and enhanced quality of life for building occupants and the community. Sustainable buildings also have productivity benefits, such as decreased absenteeism and staff turnover. The University of Florida strives to recruit and train the best staff and wants to provide an environment to retain this talent. She observed that a happy employee is often a productive employee, and this fosters staff retention. The improvements in productivity are not limited to staff, but also impact student performance, noted Armaghani.

Although building sustainably may be the right thing to do, the primary reason the university pursued this course was its commitment to education, explained Armaghani. The university wants to train leaders and encourage them to make a difference in the world. She believes it is the university's responsibility to ensure that students gain knowledge and expand this field on local and global levels. The university is empowering students to make decisions that benefit the environment and future generations.

University of Florida Sustainability Plan

Since 2001, the University of Florida has officially pursued LEED certification for all major renovations and construction projects. The university projects ranged in cost from $3 to $85 million, and this major commitment required evaluation of data and construction standards, noted Armaghani. These standards are important because there is a limited budget to maintain a building once constructed. The University of Florida was among the first institutions to require LEED accreditation of its staff. With this preparation, staff were empowered to take leadership in the building design process, to ensure that contractors and consultants understood the university's requirements, and to secure the best value for the money.

Armaghani noted that 14 of the 35 LEED-certified projects in the state of Florida are in the University of Florida system. Although the first buildings were designated for health research and medicine, newer facilities include the school of engineering, the school of law, the library, and a number of technology centers. The university has changed the construction culture on campus—suppliers, contractors, and consultants who want to work with the university must be familiar with the LEED process. The University of Florida helps shape the construction culture in the surrounding community, said Armaghani.

She observed that a number of programs facilitated this transition for the university. First, the university committed to using reclaimed water for irrigation. Second, mass transit was incorporated into the program to accommodate the large student population. Finally, the university has shown its commitment to conservation by setting aside 300 acres to remain undeveloped. These efforts help to shape future programs in sustainability, noted Armaghani.

The University of Florida has gone beyond LEED standards by committing to additional construction standards, such as Energy Star roofing, tree preservation, and waterless urinals. The installation of waterless urinals alone may save up to 40,000 gallons of water per year per urinal, said Armaghani. This is a tremendous benefit for the environment and makes economic sense, she added. Other innovations include harvesting rainwater for flushing toilets and using photosensors to determine occupancy in a room. Most raw materials used in the buildings come from within 500 miles, which reduces transportation emissions. The university is also committing to green power to reduce its environmental footprint. In a pilot project, the university started purchasing or contributing to green power for two buildings and has offset about 1.5 million pounds of CO_2 emissions. She said this is the equivalent of taking 124 cars off the road and planting 194 acres of mature trees.

> In a pilot project, the university started purchasing or contributing to green power for two buildings and has offset about 1.5 million pounds of CO_2 emissions. This is the equivalent of taking 124 cars off the road and planting 194 acres of mature trees.
>
> —*Bahar Armaghani*

Challenges of Implementation

Armaghani noted a number of internal and external challenges to the implementation of sustainable building. The internal challenge was institutional change. People accustomed to the way they performed their work needed convincing that the available data and cost benefits warranted change in procedures. The perception that LEED-certified buildings cost more money was another challenge; however, costs were reduced by the in-house LEED administration. Armaghani explained that steep learning curves caused the initial buildings to be more expensive than later projects. She added that the university learned a number of lessons along the way, and it continues to learn. A key step was gaining understanding of how to work within the sometimes arduous LEED certification process. The university assigns a LEED coordinator to oversee the project through the process. External challenges were posed by contractors, subcontractors, and consultants who were unfamiliar with the LEED process and required training.

EMORY UNIVERSITY

Wayne Alexander of Emory University observed that—much like the University of Florida—Emory University has not encountered significant conflicts with sustainable building. Because the university is an urban campus in the heart of a residential area, there is general agreement that commitments to the philosophy and concept of sustainability are appropriate not only for the good of the university proper, but also for the community at large, said Alexander. Emory has undertaken many sustainability initiatives, but these have been less formal than current plans, noted Alexander. The Emory environment was originally conceived as a forest in which students walk across campus under its canopy. Although part of the canopy has suffered over the years, there are still large areas of virgin forest on campus. The current plan aims to preserve this original concept.

> Because Emory University is an urban university in the heart of a residential area, there is general agreement that commitments to the philosophy and concept of sustainability are appropriate not only for the good of the university proper, but also for the community at large.
>
> —*Wayne Alexander*

Sustainability Vision

Alexander credited university leaders with taking a significant role in establishing the original vision and implementing the sustainability initiative at Emory. The sustainability vision was codified in a document that states, "We seek a future for Emory as an educational model for healthy living, both locally and globally—a responsive and responsible part of a life-sustaining ecosystem." From this vision, the university has focused considerable effort on human health, which is reflected in five primary themes of the vision:

- A healthy ecosystem context
- A healthy university function in the built environment
- Healthy university structures, leadership, and participation
- Healthy living, learning, and working communities
- Education and research

The university has established effective programs to meet these goals, noted Alexander. One such initiative is the Institute for Predictive Health Care—a joint project with the Georgia Institute of Technology. The institute is dedicated to maintaining health rather than simply treating disease, and it has devoted considerable time to discussing that concept. What occurs in its hospitals is ultimately defined by what is occurring in the community, observed Alexander. The university insures 10,000 direct employees as well as another 20,000 people and their families. Sustaining their health is a prime institutional consideration.

Need for a Sustainable Vision

Alexander remarked that the rise of obesity and type II diabetes is an epidemic of potentially catastrophic proportions in the United States. He described this as a problem of "too's": people eat too much food of poor quality and exercise too little. The need to address this epidemic from a preventive point of view is reflected in Emory's sustainable vision. The objective, said Alexander, is to make the campus a model for healthy living and to convey this message to anyone who visits the campus. He explained that the goal of all the education programs is to transform Emory graduates into ambassadors for sustainability and healthy living.

The case for sustainability began some years ago when Emory embarked on a LEED pilot program. The initiative was fully supported at the highest leadership levels within the university, including the board of trustees. A number of reasons for adopting LEED were articulated:

- Supporting the environmental mission of the university
- Providing the framework for high-performance buildings
- Providing third-party validation of the sustainable vision
- Making good business sense by using life-cycle analysis and not first-cost analysis to make decisions on equipment and building fixtures
- Providing leadership and educational opportunities on sustainability
- Being good stewards of the environment

LEED at Emory

The Whitehead Research Building was the first building on campus to be certified at the LEED Silver level. Coming in ahead of time and slightly under budget, it exemplifies how to use sustainable architecture to ensure health, noted Alexander. The university's commitment to LEED-certify all of its buildings is demonstrated by the 11 LEED projects currently registered at Emory. The university is approaching 2 million square feet of space that is being designed with the LEED program.

The Winship Cancer Center was the university's first experience in building a green healthcare facility, noted Alexander. As a primarily outpatient-based center, it did not pose the challenges of inpatient facilities and their attendant complexities. Alexander remarked that the building is quite appealing as a place to work, with a reliance on natural lighting.

LEED has been an effective way to analyze the environmental impacts of a project, including energy usage predictions, observed Alexander. The cost of a LEED building for Emory has been 0.5–2.0 percent above traditional approaches, and the payback has generally occurred in the first few years of a building's operation. Going beyond LEED, Emory strives to incorporate sustainability into its operational and academic endeavors. The university is continually developing both a strategic and a master plan. These plans address such issues as traffic flow problems, stormwater runoff, energy conservation and the development of alternative energy, and reduction of the university's ecological footprint. Traffic flow problems go beyond Emory's border and require partnerships with the county, neighborhood associations, the Centers for Disease Control and Prevention, and the American Cancer Society. As a result of its commitment to a holistic approach to sustainability, Emory has adopted a plan that

> Going beyond LEED, Emory strives to incorporate sustainability into its operational and academic endeavors.
>
> —*Wayne Alexander*

- calls for all facilities to be certified at the LEED Silver level at a minimum;
- is an integral part of the Emory University Sustainability Initiative;
- allows the sustainability commitment to inform planning but not limit growth;
- directs all facilities to support healthy lifestyles—not only among the ill, but also among the well who work at or visit the campus; and
- emphasizes health preservation guided by the Emory/Georgia Tech Institute for Predictive Health Care.

Because of the relatively large size of the healthcare industry in the U.S. economy, it can make a highly favorable impact on human health by minimizing its ecological footprint. This can be accomplished if health care institutions fully embrace and adopt sustainability principles in general and especially in facility construction, concluded Alexander.

6

Champions for Change

Implementing new approaches to constructing buildings requires commitment from institutional leadership. Roger Oxendale and Scott Slotterback contributed to the successful adoption of green building practices at their respective healthcare institutions. Roger Oxendale's presentation at the workshop focused on implementing the decision to build green in university medical centers, while Scott Slotterback shared his expertise on the subject of building green on a large scale.

CREATING A COMPREHENSIVE GREEN HEALTHCARE SYSTEM

In the 1940s, Pittsburgh was extremely polluted, acquiring the nickname of the Smoky City; the streetlights were turned on by 3 p.m. every day because of the darkness created by smoke from the steel mills. Local leaders in the 1940s met with the architect Frank Lloyd Wright and asked him what could be done to improve Pittsburgh—Wright's suggestion was to abandon it. The city leaders did not choose to abandon the entire city; instead, they considered how to change the environment and stimulate new ways of thinking, said Roger Oxendale of the Children's Hospital of Pittsburgh at the University of Pittsburgh Medical Center (UPMC). The leaders made an insightful move toward sustaining the long-term health of the people in the region. One of the first steps was to initiate the requirement for businesses and civic leaders to change from coal (the primary cause of pollution) to gas and other smokeless fuels for heating. This was the beginning of a significant green renaissance for the Pittsburgh region, noted Oxendale. By moving forward with courage and conviction, Pittsburgh's leaders created a livable, diverse economic region, with one of the most highly regarded and sophisticated healthcare systems in the world.

Major Components of Greening Throughout the University of Pittsburgh Medical Center

Today, it is imperative that regional leaders continue working to improve the health of residents, noted Oxendale. The University of Pittsburgh Medical Center is the largest employer in western Pennsylvania, and the second largest employer in the state of Pennsylvania. UPMC healthcare providers have a significant role in efforts to provide health information to the community.

UPMC used this role to create a vision of a comprehensive green healthcare system, one that embraces

- constructing new high-performance buildings,
- taking a leadership role in western Pennsylvania in educating families and schools about the effects people have on their environment and their own health,
- applying and sharing the scientific research the hospital pursues to advance the treatment of children,
- improving the training of medical residents in green health care, and
- incorporating green practices and treatment into the health care and the overall improvement of children, both in their homes and in their communities.

Part of the mission of the Children's Hospital of Pittsburgh is to use information to leverage its activities. The hospital's leadership goes beyond its four walls and extends into the community, said Oxendale. In conjunction with the provision of children's health care, hospital leaders consider numerous areas of importance, including how the community lives, how the hospital is operated, how research is performed, how patients are cared for, and how the community is affected by the hospital.

Two key partners for constructing the green pediatric hospital were the Heinz Endowments and the state of Pennsylvania. They awarded the Children's Hospital of Pittsburgh $5 million toward the construction of a new green hospital. UPMC is working to construct new green buildings, retrofit existing buildings, and transform health care in numerous areas, such as chemical and hazardous waste management, air quality, energy and water conservation, and housekeeping, said Oxendale.

The new hospital campus has a budget of $575 million and will be built on a 10.2-acre campus. The research building is scheduled for completion in 2008, and the hospital building will open in 2009. Approximately 60 percent of the hospital will be new construction, and the remainder will involve renovation and retrofitting of the old medical campus (Figure 6-1). The challenge is to construct a technologically advanced research building by retrofitting a partially demolished building and incorporating it with new construction. According to Oxendale, the Children's Hospital of Pittsburgh is among the first pediatric hospitals in the country to apply for LEED certification.

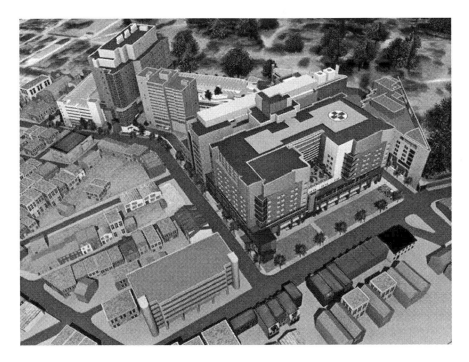

FIGURE 6-1 The new UPMC hospital campus has a budget of $575 million and will be built on a 10.2-acre campus, with 1.2 million square feet of floor space. New construction will make up about 60 percent of the hospital, and the rest is an old medical campus, which will be renovated and retrofitted.
SOURCE: Children's Hospital of Pittsburgh (2005, unpublished).

The Center for Building Performance and Diagnostics at Carnegie Mellon University is currently engaged in surveying UPMC's existing operations and practices. It has provided a set of recommendations for environmental sustainability and future innovations.

Purchasing practices are a critical area for both cost reduction and environmental gains. The Carnegie Mellon evaluation identified the purchasing of medical and laboratory supplies, patient comfort and care supplies, office supplies, and food services as areas of focus. In addition, it suggested that UPMC should identify and test alternative cleaning products and track public health and environmental benefits of alternative operations.

UPMC staff learned how to develop, test, and (when possible) implement strategies for the efficient conversion of heating, cooling, ventilation, and lighting systems. They also gained understanding of other approaches to reduce consumption of natural resources. In their effort to reduce waste streams, the hospital focused on a multiple waste management approach through recycling and disposal methods for chemotherapeutic, radioactive, and other infectious and hazardous waste products, noted Oxendale.

Reduction of vehicle pollution was also taken in consideration. The hospital will use hybrid or nondiesel parking shuttles and public transportation. Employees are encouraged to carpool and drive alternative fuel vehicles. Other strategies currently being evaluated and planned include the provision of bike racks and showers and electric outlets for hybrid cars, said Oxendale. The hospital also plans to ensure that truck engines are shut down at loading docks during deliveries.

The impacts of building design on healthcare outcomes, employees, and the work environment are being studied at the Center for Building Performance and Diagnostics and the University of Pittsburgh's Mascaro Sustainability Initiative Center. Healthful initiatives, such as access to open areas, walking paths, and ergonomically designed furniture, will be incorporated. New patient rooms are designed with green materials and additional space in which patients and families have privacy and some level of control over their environment. The hospital also plans to establish a rooftop healing garden to promote the healing environment.

UPMC's Collaboration Efforts

The University of Pittsburgh Cancer Institute, the Graduate School of Public Health, and the Center of Environmental Oncology formed a team to work on the elimination of persistent toxins, such as mercury. The hospital also plans to reduce the use of toxic chemicals, reduce the quantity of water used for cleaning, and use new materials to improve infection control. For example, the hospital is currently implementing the use of microfiber mops. In addition to any health benefits, these measures save thousands of gallons of water a year and require fewer cleaning chemicals, noted Oxendale.

Among other environmental initiatives, the team is looking for ways to buy equipment and medical instruments that are state of the art and less harmful to the environment. One focus area is the recycling and donation of surplus hospital supplies to developing nations. In addition, use of the electronic health record is being implemented throughout the Children's Hospital of Pittsburgh. As a result, most filing space and cabinets for paper records are being eliminated, and it is anticipated that the new campus will be paperless.

With the University of Pittsburgh School of Engineering, UPMC is working to optimize software and control systems that will reduce costs for chilled water, steam, electricity, and water disposal throughout the system. These significant measures will enable the development of baseline data to be used in tracking the impact of the new green Children's Hospital of Pittsburgh. The hospital will be a highly efficient, sustainable facility—it will incorporate water and energy conservation, improved air quality, green building materials and cleaning processes, and light, open, airy spaces, said Oxendale.

The leadership measures taken to clean up the Pittsburgh region almost 60 years ago laid the groundwork for a future in which hospitals can be models of disease prevention and cure. The greening of the pediatric institution will allow the medical community to continue improving the futures of their young patients, concluded Oxendale.

BUILDING GREEN ON A LARGE SCALE

Media commentator and author Marshall McLuhan once said: "If we drove the way we typically plan, we would spend most of our time looking into our rearview mirrors and we would crash our cars." We often look to the past so we can build on our traditions and learn from our mistakes, said Scott Slotterback of Kaiser Permanente. Today, we need to look to the future, project where we want to be, and then determine how to get there. This is especially true when trying to build green on a large scale. Kaiser Permanente is using *The Green Guide for Health Care* to define what is green and as a foundation for the next version of the *Eco-Toolkit* (a design and construction resource guide), said Slotterback.

Located in nine states and the District of Columbia, Kaiser Permanente is one of the nation's largest nonprofit health plans. It has 8.3 million members and 60 million square feet of occupied floor space in more than 900 buildings. Kaiser Permanente is an integrated model of care, consisting of three companies: the Kaiser Foundation Health Plan (an insurance company); Kaiser Foundation Hospitals (a healthcare provider); and the Permanente Medical Groups (a for-profit group of physicians). Kaiser Permanente's planned facility growth includes building 23 new hospitals (6 of them on green-field sites), major hospital bed expansions, replacement or significant addition projects, thousands of renovations, new medical office buildings to support hospital projects, as well as additional utility plants and parking structures, noted Slotterback.

Kaiser Permanente established an Environmental Stewardship Council to lead the effort in going green on a large scale. The Environmental Stewardship Council reports directly to the chief executive officer and is driven by a vision statement, stating that Kaiser Permanente "aspires to provide healthcare service in a manner that protects and enhances the environment and the health of the community now and for future generations" (Slotterback, 2006). The company's role is not limited to protecting people in the buildings, but also involves making decisions that protect the communities in which members live, noted Slotterback.

Formed in 1998, the Kaiser Permanente Alliance Program is a group of architects and contractors serving the company's design and construction needs. The company brought its physicians, nurses, environmental health specialists, and individuals responsible for maintaining and operating its buildings into discussions at the very beginning of the construction project. This integrated design approach ensured that all groups involved were educated in the same concepts.

Among other initiatives, pervious paving to filter water from parking lots back into the aquifer was introduced. The cost of the pervious paving per square foot was higher than the cost of conventional asphalt, said Slotterback. However, from a system standpoint, conventional asphalt would have required a connection to the city's storm sewer system. Because all of the pipelines to that system were not nearby, the cost of installing large piping might have cost more than the pervious paving. Other Kaiser Permanente initiatives include using drought-tolerant native species in landscaping (they help reduce water consumption and costs in maintenance), providing access to daylight and views in the facilities, installation of photovoltaics in lieu of mechanical screens on one portion of the building, taking steps to eliminate polyvinyl chloride (PVC) in the building materials, using recycled content in materials, encouraging alternative transportation, and creating tobacco-free sites.

Standards for Kaiser Permanente Design and Construction

As a large-scale consumer, Kaiser Permanente is able to leverage its buying power to transcend institutional barriers, said Slotterback. Taking advantage of the company's large scale, Kaiser Permanente was able to create incentives that transformed not only what was available to them, but also the marketplace at large.

Kaiser Permanente's National Facility Services have responsibility for the design and construction of the new facilities. A robust standards program monitors quality control and facilitates design and operational efficiency; it is also used to promote the company's green building program.

Kaiser Permanente's National Purchasing Agreement program (NPA) joins with manufacturers to drive down costs, provide equipment with longer operating lives, and support the company's green agenda. Compliance with the NPA is mandatory for the company's design and construction teams, and this strategic alliance enables Kaiser Permanente to develop products and systems that meet their specific needs, noted Slotterback.

In 1993, Kaiser Permanente negotiated the first NPA for carpeting. The request for proposal included a requirement for bidders to state how they reduce waste and support recycling. Most of the manufacturers did not understand, and only one manufacturer indicated that they were recycling carpet in response to Kaiser Permanente's request, said Slotterback.

In 2002, the carpet contracts for the NPA were up for renewal. This time, Kaiser Permanente decided to focus on sustainability of current and potential partners, explained Slotterback. The company's negotiating team included interior designers, a representative from the janitorial group, and the director of environmental stewardship. In addition, two other members of the green buildings committee, an outside architect, and a representative of the Healthy Building Network were included in the team. The team was charged with focusing on three main criteria in evaluating current and potential builders: sustainability, product performance, and aesthetics.

The negotiating team conducted research on the carpet industry to identify which companies were leaders in sustainability. To gain understanding of the environmental impact of carpet fiber, the team also met with fiber manufacturers. Kaiser Permanente partnered with the Healthy Building Network to develop a detailed questionnaire for determining the chemical composition of carpets and the environmental impact of carpet manufacturing. The questionnaire was used to assess effects on the environmental quality of buildings and their surrounding communities. Because conventional testing was not adequate for Kaiser Permanente's needs, the company developed a special testing procedure to measure impermeability.

After deliberations, the team selected five carpet companies; two that were currently under contract and three that were included on the basis of their leadership in sustainable practice. The negotiating team then prepared a request for proposal and sent it to all five companies. The team invited each manufacturer to make a presentation with a focus on their sustainable practices, healthcare product line, and product performance. The team then met and scored each company on the basis of the selection criteria.

Sustainable practices and product evaluation or performance were assigned a weight of 45 percent each, and green innovation was assigned a 10 percent weight. An ideal product did not exist at that time, noted Slotterback. A major issue for Kaiser Permanente was the elimination of PVC, and all products meeting the performance criteria had PVC backing in the carpet. The two companies selected, Interface and C&A Floorcoverings, had aggressive recycling programs and agreed to work with Kaiser Permanente to develop a PVC-free carpet backing that met performance criteria in a 2-year time frame.

C&A Floorcoverings managed to meet the criteria, and they have been awarded Kaiser Permanente's NPA for carpet in all of Kaiser Permanente's facilities. C&A Floorcoverings developed a carpet backing that performs at the level of PVC but uses material that is reclaimed from laminated safety glass and contains 96 percent postconsumer recycled content.

Kaiser Permanente explored the environmental performance of other flooring products as well. The company looked for a replacement of dominant surfaces, such as vinyl sheet flooring and vinyl tile flooring, within the hospital. Kaiser Permanente revised its standard to use two alternative flooring systems: Stratica by Amtico and Nora rubber flooring, noted Slotterback.

Chemical content of the materials is not the only important component of Kaiser Permanente's flooring standards, said Slotterback. The maintenance process is very important as well. The maintenance cycle of PVC-based products usually requires stripping and waxing. The stripping releases particulates that may trigger asthma; in addition, the cleaning chemicals used to treat PVC-based products are believed to be harmful. The Nora and the Stratica flooring requires

minor cleaning (mostly damp mopping), and lower VOC-emitting cleaning sol-
vents are used for more thorough cleaning.

Slotterback concluded that additional research would benefit consumers and
give a firmer foundation for Kaiser Permanente's programs, such as the PVC
removal program. To make intelligent consumer decisions, a better understanding is needed of the impacts of chemicals used in building products, asserted
Slotterback. A federal standard—similar to the food labeling standard—that indi-
cates the content of building and furnishings products, would enable consumers
to make intelligent decisions. The ability to make informed decisions and choices
would shift the marketplace to the use of sustainable products and in turn will
help to achieve sustainability goals, concluded Slotterback.

7

General Workshop Discussion

During the general discussion, participants expanded on the presentations and the overaching scientific principles of green buildings. Russell Perry of the Smith Group facilitated the discussion by asking participants if they agreed that the implicit question in the presentations was "Can we improve human health or speed the process of healing by building higher-quality buildings?" If so, can science measure that incremental improvement and disaggregate the contributing factors to understand the effectiveness of various strategies?

WHAT IS GREEN BUILDING?

During the workshop, considerable time was spent discussing the definition of *green*. For example, one participant questioned whether green includes daylighting or noise issues. He challenged the group to define "green" more narrowly around environmental issues, including products and materials used in the healthcare environment, and sustainability. He further suggested that people need to agree that green is more than just energy efficiency. Perry expanded this discussion by drawing from Jason McLennan's book, *The Philosophy of Sustainable Design* (2004). He said that the largest difference between the environmental building movement of 30 years ago and the past 10 years is the concern about quality. The focus is now on quality environments—indoor environmental quality, water quality, the quality of light, and quality of life—and not just about energy. He credited leaders in sustainability with expanding the definition. Although participants generally agreed on the core elements (e.g., energy, low-emitting materials), the boundaries of building green were not firmly established during the discussion.

ADVANCING GREEN BUILDING RESEARCH

Judith Heerwagen noted that current research on green buildings happens in a stovepipe fashion. For example, people who study noise typically do not study other dimensions, such as lighting. It is the intersection of such lines of research that is likely to be very important. She further noted that health has multiple determinants, some in the social environment, others in the physical environment. She pointed out that when people move into a new building, there are the placebo and Hawthorne effects to consider. Most research to ascertain the effects of a building is performed once the individuals have adapted. She questioned if there is valuable information in this phase of adapting that warrants additional research.

Craig Zimring observed that hospitals have research advantages that should be capitalized on. First, hospitals measure a multitude of outcomes that most other buildings owners do not. They routinely measure patient satisfaction, family satisfaction, rates of infection, error rates, and staff turnover. These measures are often made at the unit level, which facilitates comparisons between units of different design. Second, there is centralization in the healthcare sector. For example, large hospital chains and healthcare design firms can facilitate data collection and provide the opportunity to do controlled studies. Third, there is an opportunity to engage academic medical centers. It is ironic that innovation has been driven by suburban hospitals of medium size and not the large academic health centers.

TRANSLATION OF RESEARCH INTO PRACTICE

Science has accumulated a body of literature to support green buildings, but the information is not complete at this time. Some participants observed that this is the challenge currently facing the sustainability field. Anthony Bernheim remarked that there is some uncertainty in the information, and that the precautionary principal is often used. In many ways, we know enough to be concerned but not enough to make the final judgment on many of the attributes of building green, he noted.

> There is some uncertainty in the information, and the precautionary principal is often used. In many ways, we know enough to be concerned but not enough to make the final judgment on many of the attributes of building green.
>
> —Anthony Bernheim

Concurring, Zimring elaborated by saying that currently available evidence does not yet support major design decisions. In other words, researchers can only state in general terms that building better buildings can improve outcomes. Researchers can support the concept of green design as a path to quality of care, environmental performance, or improved health, but they can-

not specify necessary individual elements. Similarly, researchers can say
that what is needed is more light or
larger patient rooms, but they cannot say how big those rooms need
to be for a given patient. They cannot say how much space should be
devoted to a restorative garden. Currently, clear data that would support
a business case for gardens are not

> Researchers need to go beyond a simple
> dichotomy of bad and good. The research
> needs to define how these individual
> elements operate in combination and
> nuanced ways.
>
> —*Craig Zimring*

available. The research agenda needs to be enhanced in this area, noted Zimring.
He asserted that researchers need to go beyond a simple dichotomy of bad and
good. Research needs to define how individual elements operate in combination
and nuanced ways.

While one participant agreed with the above position, he suggested that
defining enough evidence depends on the audience. For example, evidence that
suffices for the public may not be acceptable to the scientific community. Howard
Frumkin expanded this idea by observing that there is both an advocacy position and a research position. Individuals who are advocates may conclude that
science has amassed sufficient information to guide healthcare facilities in how
to implement green building practice. However, a researcher might note how
much is unknown about the effects of building green on human health. Although
definitions and evidentiary standards are well understood in other parts of environmental health, for green buildings some of these issues remain unsettled,
noted Frumkin.

People on the advocacy side should be straightforward about what is known
and where evidence is lacking within the field, Frumkin continued. From these
discussions, the scientific community
needs to frame data needs accordingly,
he asserted. On the research side,
the field needs to be very strategic.
Researchers do not need to conduct
research that simply answers questions for which answers are already
available. The questions that need to
be answered should guide designers and builders in what to do. The
research agenda that science needs to
develop should be very applied, very

> The research agenda that scientists need
> to develop ought to be very applied, very
> strategic, and very targeted. It should
> supplement the advocacy agenda by filling
> in holes, and allow us to move forward and
> fuse the two agendas.
>
> —*Howard Frumkin*

strategic, and very targeted. It should supplement the advocacy agenda by filling
in holes, allowing progress in fusing the two agendas, noted Frumkin. Public
health advocacy is necessary, but it needs to be based on data. One fundamental
way to fuse the agendas is the precautionary principle. It is also necessary to
identify gaps in current knowledge and to direct research accordingly.

ADDRESSING GAPS IN KNOWLEDGE

Several participants initiated a discussion about research gaps. Perry empha-sized the importance of not researching issues that are already well understood, but rather to disseminate existing information through education or publication. He noted that much research is unknown to most individuals in the architectural world and that education is an important part of solving this problem. Bernheim continued this discussion by observing that midlevel architectural and engineer-ing professionals are important decision makers and should be informed about current research.

Another participant remarked that Kaiser Permanente and other hospitals have demonstrated not only that the changes being made are cost-effective, but also that they improve quality. However, she also suggested that it is important to disseminate results from pilot studies, which can help to guide future controlled studies.

ECONOMICS

Some of the discussion followed up on the economics of research. Perry reiterated concerns about identifying the most important benefits. He questioned how facility managers know where to spend money and if there is sufficient evidence of effectiveness to justify additional expense. Another participant sug-gested that the fundamental question of how much a hospital should cost should be addressed, rather than celebrating the fact that these buildings can be con-structed for a premium of less than 2 percent of the cost of traditional buildings. John Poretto noted that the basis of the problems is separating capital costs and operating costs into individual silos. Generally, capital costs come from the sur-plus made from operations. If a building owner skimps on the capital side, it will ultimately be made up through the operational side in the indirect costs recovery scheme, noted Poretto. Perry concluded this part of the discussion by suggesting that people who are taking the greatest financial risk are those who are not follow-ing these strategies and not constructing healthy and energy-efficient buildings.

COMMUNITY LEADERSHIP

One participant said that a hospital ought to be a beacon of light for good environmental practice. Hospital and healthcare professionals often have a larger role in the community by providing credible information about public health issues. Perry agreed, adding that credit should be given to individuals who are showing leadership and giving credence to the best practices. As the field moves forward, scientists should consider what data are needed to document the importance of the green healthcare movement. A participant recommended that healthcare facilities provide better choices about green products and chemical use to communities.

Perry concluded the general discussion by saying that the meeting should celebrate the research and data that has already been compiled, as well as the leadership of companies that provide facilities to further study best practices.

Presentation Abstracts*

THE CONFLICT BETWEEN GROWTH AND GOING GREEN: THE EXPERIENCE AT EMORY

R. Wayne Alexander M.D., Ph.D.

Emory University has broadly embraced the principles and practice of sustainability, which is recognized in the university strategic plan. The sustainability vision was developed in the context of the strategic plan implementation and summarizes the goal that: "We seek a future for Emory as an educational model for healthy living, both locally and globally—a responsive and responsible part of a life-sustaining ecosystem" (Sustainability Commitee, 2005) The primary themes of the sustainability vision are a healthy ecosystem context; healthy university function in the built environment; healthy university structures, leadership, and participation; healthy living, learning, and working communities; and education and research. Emory has initiated a plan for realizing a "sustainable architecture for health." There are currently 11 Leadership in Energy and Environmental Design (LEED)-registered projects at Emory. The first LEED building in the Woodruff Health Sciences Center was the 321,000 sq. ft. Whitehead Biomedical Research Building. This building was the LEED pilot project at Emory. It was highly successful, LEED Silver certified, and came online ahead of schedule and under budget. The LEED concept has been supported by the board of trustees. The first healthcare building was the Winship Cancer Institute, which is LEED registered. Plans are for all future construction of major buildings to be LEED registered, with the goal of reaching Silver certification for all construction at the very least. These standards are to be applied to the new Emory University Hospital and the Emory Clinic buildings, which are in the planning stages.

---

*This chapter contains individually authored abstracts that were submitted to the roundtable by presenters prior to the workshop.

Justifications for the university's commitment to the LEED program include the following:

- It supports the environmental mission.
- It provides the framework to build high-performance buildings.
- It provides third-party validation of the sustainability vision.
- It makes good business sense (use life-cycle cost analysis, not first cost, to make decisions on equipment and building features).
- It supports Emory's desire to be leaders in sustainability initiatives and in stewardship of the environment.

Emory's facility development program is an integral part of the overall sustainability initiative. The commitment to this initiative to date has not limited growth but has powerfully informed planning. Programmatically, all facilities will support healthy lifestyles, not only for the ill but also for the well who work or study at, or visit, the university. The general emphasis on health preservation will be guided by the Emory/Georgia Tech Institute for Predictive Health Care.

FRAMING THE PROCESS: INSTITUTIONAL CHANGE TO GREENING A CAMPUS: SUSTAINABLE CONSTRUCTION AND BUILT ENVIRONMENT AT THE UNIVERSITY OF FLORIDA

Bahar Armaghani, B.S., LEED AP

The University of Florida's Facilities, Planning and Construction Division (FP&C) is committed to developing a sustainable campus and delivering sustainable buildings to the University of Florida (UF) in support of maximizing efficiency, productivity, and good health and comfort of the faculty, staff, and students.

The University of Florida was thinking green and testing green before green practices were even on the radar for most educational institutions. In the late 1990s, sustainable design and green building concepts were being tested on several new projects. In 2000, sustainable design elements were incorporated into the UF master plan and construction program documents. In 2001, FP&C adopted LEED criteria for design and construction of all major new construction and renovation projects. The UF faculty committees followed this effort with full endorsement. In 2005, FP&C raised the bar on this arena and established a minimum goal of silver LEED certification for all university projects.

The University of Florida has made significant strides toward the goal of being a leader in sustainable development and incorporated this into the UF fabric to serve the interest of the students, staff, faculty, our community, and the world. We were proactive in taking this posture and adopted LEED when it was at its infancy in support of building a healthy environment on campus.

Since 2000, FP&C has achieved the following milestones:

- LEED-certified buildings (totalling 79,107 GSF) including:
  - Rinker Hall—LEED Gold certified
  - McGuire Center for Lepidoptera and Biodiversity Research (butterfly museum)—certified
- LEED-registered buildings in design and construction phase (totalling 1.1 million GSF) including:
  - Cancer and Genetics Research Center Pavilion
  - Orthopedic Surgery and Sports Medicine Institute
  - Shands Biomedical Research Laboratory
  - Nanoscale Institute Research Facility
  - Food Animal Veterinary Medicine Facility
  - Powell Structures and Materials Laboratory
  - Legal information and phase II law building
  - Library West addition and renovation
  - Baseball locker room facility
  - Mary Ann Cofrin-Harn Pavilion (museum)
  - Hub renovation (technology center)

We have enhanced the construction standards to incorporate LEED criteria and have raised the bar in delivering a healthy building environment. The unique and challenging aspect of the green buildings on our campus is that every building is different in size and function. Also, the university's FP&C has taken the lead to work with Shands Hospital on their new hospital construction to bring the hospital component into sustainable design. The success of building green on campus has generated a ripple effect throughout the campus manifesting in a desire to look into other sustainable practices such as zero waste by 2015, reducing carbon emission, and green purchasing. These are a few of the new initiatives that the university president announced last October on Campus Sustainability Day.

The University of Florida is leading our state in the design and construction of green buildings. This has been made possible by the support of the university administration, the faculty senate, and the tremendous enthusiasm of the staff, faculty, and students.

Earlier green practices have played an important role in creating a sustainable campus including

- converting campus-wide irrigation to use reclaimed water generated by the UF-run water reclamation facility that processes over 2 million gallons of reclaimed water per day,
  - a mass transit system,
  - a no smoking policy,
  - maintaining over 300 acres of conservation land,

- a full recycling program,
- commissioning, and
- an indoor air quality program.

We have come a long way, but we know that we have a long way to go. Our green building approach has evolved and expanded from using LEED for new construction (LEED-NC) to using LEED for existing buildings (LEED-EB) and for health facilities. Over the years, our commitment has strengthened, and our enthusiasm has grown to build more sustainable and healthy buildings. With this commitment, we strive to include our campus community and other surrounding communities in this process. We involve our students in the process and teach them unforgettable hands-on lessons. When they graduate, they will be prepared to make the right decisions as consumers and conservers toward saving the environment.

BUILDING GREEN AND INTEGRATING NATURE: RIKSHOSPITALET UNIVERSITY, OSLO, CASE STUDY

Knut H. Bergsland

This case study was presented because of its qualities in terms of humanizing the hospital environment, integrating nature, and giving access to direct daylight to all patient rooms and most of the functional working spaces. Natural materials were utilized as far as possible according to the LEED-NC version 2.2 registered-project checklist, as such Rikshospitalet would probably achieve certification.

Building Green

Regardless of the scope of the definition of *green building*, it is imperative to seek the most important indicators in terms of individual, environmental, and community health. Green building must include a vast array of subjects. Still, there is a need to pinpoint the most important indicators, the ones that most benefit the health of the patients and personnel with the least effort and use of resources. In terms of hospital operating, it is imperative to establish a committed culture for operating and maintaining a sustainable building concept, including all its support systems throughout the entire life cycle. What is needed is a hospital concept for maximum, long-term performance on the most important indicators.

Integrating Nature

The importance of nature as a stress-reducing trigger for the healing process has been an established fact for quite a long time. To take a few shortcuts, it may follow from this that planning for maximum daylight and integrating nature in the hospital concept by as many means as possible is a right thing to do in both the patient and work environments. Seeking the most crucial elements in terms of health return (environmental and medical outcome) is important also in this respect.

Background to the Case Study

Norway spends 10 percent of its gross domestic product on health care (Johnsen, 2006), as opposed to the 16 percent in the United States (CIA, 2007). The health-care system is 90 percent public and tax based; hospital inpatients do not pay for their stay (Bergsland, 2005). Hospitals are owned by the state, but they are run as trusts. Competition between hospitals was introduced a few years ago, and doctors are employed by the hospital. The Norwegian healthcare system is driven by the same forces as most other Western countries—demographic change, technology, and demands for efficiency; but the system is still run within the framework of a national healthcare system based on equal access to and distribution of services as the main principle.

Rikshospitalet University Hospital

Rikshospitalet—built on a virgin site just outside the city center—is a tertiary teaching and referral hospital, and covers all clinical specialties, except for geriatrics and psychiatry. The 1,233,000 sq. ft. building, completed in 2001, has 585 beds, excluding intensive care. There are 35,000 inpatients, 20,000 day patients, and 160,000 outpatients per year, with a workforce of 4,000 full-time equivalent (FTE) positions. A substantial clinical production growth from 2001 to 2004 has been absorbed by the building; however, this has not occurred without straining the ventilation and energy systems. Productivity levels are up more than the 15 percent above the rise in staffing levels. Absenteeism decreased from 8 percent to 6 percent.

The location of the site was chosen by the Norwegian Parliament. The hospital was built on cultivated land, despite protests from environmental activists. The site itself is sloping and saucer shaped, which was utilized by the architect to make the 5- to 6-story building appear as a nonfrightening 3- to 4-story set of buildings.

Village Structure

Rikshospitalet is conceived as a village structure, with a main square and a landmark tower, a street hierarchy and separate, but interconnected buildings. The dominant, slightly curved, 280-meter long circulation artery has a glass roof, which lets daylight into a bigger proportion of indoor spaces than in similar covered spaces. Glimpses of nature, plus sculptures and other art objects, aid wayfinding by making it easy to draw a mental picture of the route to one's destination. The curvature hides the length of the corridor, gives no long drab vistas, and reduces the need for signage. The art and glimpses of nature at intersections helps one remember and aids recognition, which facilitates patients' and relatives' trip to their destination.

Stress-Reducing Qualities

The circulation artery, with its dense pedestrian traffic, integrated art, frequent art exhibitions and concerts, and access to a grand piano—also for patients, obviously fills one important requirement for stress-reducing factors in hospital environments (Ulrich, 1991):

- A place for positive distractions in physical surroundings
- Access to social support
- A sense of control with respect to physical and social surroundings

The low, nonfrightening appearance of the building volumes and frequent access to nature and daylight may contribute to a sense of control in patients and visitors. There have, however, been no studies so far to confirm this. Art is integrated in the building. Nine percent of the total building budget was earmarked for art in the hospital. One may ask whether art as a background for activity can have similar effects as nature on stress reduction and healing. Some effects of pictures of nature and smiling human faces on stress reduction in patients have been documented (Ulrich, 1991). In Rikshospitalet, such pictures are not much used in patient areas.

Daylight in as many spaces as possible is a positive contribution to staff well-being, according to a preliminary study on the effects of the building concept on activity and productivity (Bergsland, 2005). On the other hand, daylight requirements result in longer walking distances, more circulation space, slightly lower space efficiency, and higher energy needs.

Daylight vs. Energy Use

The glass roof brings ever-changing daylight into the main street. It could be called a lovely energy drain, as the street is kept at a temperature of 17°C during the winter season. This requires extra heating, which, however, is more than outweighed by the positive effects on staff morale. The hospital's technical systems still need some upgrading. But to introduce such systems visibly in the main street volume was flatly rejected.

Rikshospitalet uses more energy per square meter than most other Norwegian hospitals, a little more than 400 kWh per square meter per year, versus under 200 kWh in new hospital projects (energy use is calculated as energy supplied by the outer wall of the building per intentionally heated area). Still there has been a more than 10 percent reduction in total energy use from 2002 to 2004, even with a substantial growth in clinical production. The hospital administration has committed itself to an ambitious program of saving energy.

Nature's Materials

Norwegians love nature's materials—especially wood. Rikshospitalet is showing the patient respect through the use of high-quality, lasting materials. Natural stone is used in the floor of the main street, on some other floors, and in street furniture. Wood is used for benches, chairs, reception desks, and in special rooms, such as libraries and auditoria. Cafeteria and other common rooms frequently have parquet flooring. Trees are incorporated in some indoor spaces and may aid biofiltration of indoor air.

Integrating Nature in Practice

The virgin site location is the major reason for the ability to integrate nature and daylight in the project: from the use of the surrounding woods for activities, access to (most) courtyards, glimpses of nature at intersections, to the preservation of existing, big trees, and so on. The trees also play a role in achieving a human scale in the project.

Partly Green and Integrating Nature

In terms of the LEED checklist version 2.2. Rikshospitalet seems to meet some of the criteria for sustainable sites, but not all.

The hospital's strongest points seem to be

• daylight to as many spaces as possible, worth both the extra first cost and the extra operating costs—and a key to achieving the humanistic goals of the project;

• the village main street creates a place with identity and interest, generating a sense of high quality, without showing off; and

• the seemingly low building counters the impression of the hospital as a big, clinical machine.

The architects' strong will, empathy, and commitment to human values seem to be the reasons behind the success of the project as a healthcare setting. In terms of green building, there are still goals to be achieved.

THE CASE FOR GREEN BUILDINGS II:
HEALTH DESIGN PRINCIPLES IN HEALTHY BUILDING

Anthony Bernheim FAIA, LEED AP

Global and Local Ecological Health

Life on earth is dependant on clean air, fresh water, biological diversity, and healthy soil (for growing food and, more recently, the raw materials for rapidly renewable building materials). Because the way we design, construct, and operate buildings has a major impact on the earth's environment, we need to focus our attention on sustainable, green, and high-performance building as a way to ensure that future generations may also enjoy equal or improved health and environmental benefits.

When we think of green building, we generally think about energy efficiency and the U.S. Green Building Council's (USGBC) LEED green building rating system (USGB, 2006). However, sustainable, green, and high-performance buildings are much more complicated than this. They involve an integrated approach to energy conservation and efficiency; indoor environmental and air quality; and the efficient, effective use of site, water, and material resources. Genuine long-term environmental sustainability means more than the mainstream construction of buildings according to outdated conventions. It entails designing and constructing deep green "restorative" buildings, those that enhance the environment by producing more energy than they consume, and those that provide comfortable indoor environments with healthy indoor air quality (IAQ) (McLennan, 2004). These restorative buildings support and promote improved occupant health and reside at the highest level of the "green thermometer," a relative measure of both a building's environmental sustainability and its contributions to its occupants' physical well-being.

Health in Buildings

Because we breathe without conscious effort, we spend little time thinking about what enters our systems with those breaths. We do not see, and only sometimes smell, the chemicals and particulates that endanger our health. Yet indoor air quality is not a primary focus of contemporary building design. The U.S. Environmental Protection Agency (EPA) estimates that Americans spend almost 89 percent of their time indoors (at home and at work), 6 percent in vehicles, and only about 5 percent outdoors. They further tell us that the air indoors is about 2 to 5 times more concentrated with chemical pollutants than the air outdoors, with the result that we are being exposed to high levels of chemical concentrations for the vast majority of our lives. Our bodies, not designed for this, are responding with health afflictions such as

- sick building syndrome (short-term health effects with coldlike symptoms that can not be traced to specific pollutant sources),
- building-related illnesses (diagnosable illness whose symptoms can be identified and whose cause can be directly attributed to airborne building pollutants), and
- multiple chemical sensitivity (a condition in which a person reports sensitivity or intolerance to a number of chemicals and other irritants at very low concentrations).

Indoor air quality is dependent on a number of factors, including the quality of the outside air that we bring into the building; the chemical emissions from the materials, furnishing, and equipment that we place in our buildings; the efficacy of the ventilation systems that we use to purge the indoor air; the activities of the building occupants; and the long-term maintenance of the buildings and their contents. These factors contribute volatile organic compounds; microbial organisms and microbial volatile organic compounds from mold; semivolatile organic compounds from fire retardants, pesticides and plasticizers; inorganic chemicals such as carbon monoxide, nitrogen dioxide, and ozone; and particulate matter generated outdoors by fuel combustions and indoors by occupant activities and equipment.

Four Principles of Good Indoor Air Quality Design

In the early 1990s, my firm began an earnest exploration of the role of design in improving indoor air quality. Our work was influenced by and tested during a major civic project, the San Francisco Main Library. Through extensive research, analysis, and real-life applications, we concluded that building owners, operators, architects, interior designers, and engineers can have a major impact on a building's indoor air quality. Our experiences with that project and numerous others since then have confirmed that healthier buildings result from the adherence to four basic principles:

- Source control (reducing the indoor chemical concentrations by reducing or eliminating the pollutant source)
- Ventilation control (providing adequate ventilation to dissipate and purge the indoor air pollutants)
- Building and IAQ commissioning (a process used to check and verify that the building is constructed as designed and operates as intended)
- Building maintenance (regular inspection, maintenance, and cleaning of the building and its contents)

From Science to Practice: Source Control

There have been many developments in the science and practical applications leading to improved indoor air quality. Most recently, those developments have been in the area of source control, the principle on which I will focus in this article. Significant scientific research has been published in the area of source control and the reduction of potentially harmful substances in indoor air. Although more research is needed to build on the current body of IAQ knowledge, the collective data has provided some guidance to building designers that, combined with practical building experience over the last 20 years, has led to the current state of IAQ knowledge.

Beginning in the early 1980s, the American Society of Heating, Refrigerating, and Air-Conditioning Engineers (ASHRAE) developed IAQ guidelines as a rule of thumb limiting building occupants' long-term exposure to a small percentage of the occupational exposure (one-tenth the threshold limit value). Little was known at the time about the effectiveness of this guideline, and concerns were raised regarding the factor of safety of the indoor air chemical concentrations. Lars Mølhave of Denmark developed a total volatile organic compound (TVOC) approach to selecting indoor building materials based on the odor, irritation, memory, task performance, and other effects of these chemicals on the building occupants (Levin, 1998). An early application of this work took place in the state of Washington's East Campus building projects (Black et al., 1993). Concerns were still raised, however, about the health impacts of individual chemicals and the synergistic effects of a combination of chemicals in the air.

In 1989 my firm was selected as part of a large team to design the new 381,000-sq.-ft. San Francisco Main Library. We were concerned about the building's health impact on the library staff and patrons and incorporated IAQ into the project design criteria. We developed specifications limiting the emission of a few volatile organic compounds that were known to be odorous and have some health impacts, and we selected the building materials based on a careful analysis of technical data provided to us by the materials' manufacturers.

The most important information that we requested and eventually obtained for analysis was the chemical emissions test reports that provided us with data on each material's TVOC emissions and some of the individual volatile organic compound emissions (Bernheim and Levin, 1997). Although the library staff was originally skeptical that we could design for good indoor air quality, the building opened in April 1996 with very positive response from the staff about the IAQ. The unfortunate lesson that we learned on this project was that, although we were able to have material manufacturers eliminate some odorous and potentially harmful chemical emissions from their products, they replaced them with others about which the health effects were less well established.

By 1999, work had begun on the design of a 479,000-sq.-ft. California State office building located in the Capitol Area East End Complex of Sacramento, to be occupied by the Department of Education. An engineer working in the State

Health Department and a national IAQ expert developed a procurement specification for the building's furniture, which was intended to help the state acquire large quantities of office systems furniture with high-recycled content and low individual chemical emissions.

My firm was selected to join the team that would design and build the project. We formed a green team (including the national IAQ expert, Hal Levin of the Building Ecology Research Group) within the larger team to enhance the project's sustainability and long-term performance. We were requested by the state to give particular attention to delivering a building through the design-build process with good IAQ. We built on the previously prepared furniture procurement specifications and subsequently adapted their methodology for the building materials (Bernheim et al., 2002). The goal was to reduce indoor chemical concentrations by reducing or eliminating chemicals of concern that are carcinogens, reproductive toxicants, and chemicals with long-term or chronic health effects.

To do this, we needed to better understand the contribution of these materials to overall indoor chemical concentration and the potential health impacts of these concentrations. The California Office of Environmental Health Hazard Assessment (OEHHA) has developed a list of about 80 chemical compounds and has, through evaluation of the available science, determined the impact on the human body of long-term exposure to these chemicals. It has further developed a chronic reference exposure level (CREL) for each chemical, which is the concentration or dose "at or below which adverse health effects are not likely to occur from a chronic exposure to hazardous airborne substances. They are intended to protect individuals from chemical injury, including sensitive sub-populations" (Alexeeff et al., 2000).

Our team developed a special environmental requirements construction specification, now known as section 01350, for this project. This specification requires chemical emission testing for interior materials and sets maximum chemical concentrations based on the OEHHA CRELs, minimum material recycled content based on the State Agency Buy Recycled Campaign (SABRC), and procedures for dealing with mold on the construction site. Section 01350 also establishes an airing out period prior to substantial completion. Postoccupancy air testing in the Capitol Area East End Complex was performed, and the results indicated that the section 01350 material testing was effective in limiting the chemical concentrations in the completed building, which achieved a USGBC LEED gold rating.

Market Transformation

As design for healthy indoor air quality gains a foothold, these early projects are becoming a baseline for standards that are being followed in many industries. Section 01350 has now been incorporated into the California Department of General Service's standards for all future state buildings. My firm is incorporating these specification requirements into several of our upcoming projects, including a medical office building for the University of California, San Francisco; the Osher Center for Integrative Medicine; a large San Francisco hospital complex; and a new, 1.5 million-sq.-ft. State office building, the West End Office Complex. The guidelines have been incorporated into the California Collaborative for High-Performance Schools program, and they are referenced in the *Green Guidelines for Health Care* and the USGBC LEED green building rating system for new construction, version 2.2.

Numerous building products, including ceiling tiles and floor materials, have been reformulated by their manufacturers to reduce chemical emissions based on these specifications, and more recently, many industry trade groups have developed or are in the process of developing certifications to indicate some level of compliance. Examples include the Carpet and Rug Institute's Green Label Plus program and the Resilient Flooring Institute's FloorScore Seal (CRI, 2004; SCS, 2005). Recently, the GreenGuard Environmental Institute introduced its Standard for Children and Schools, which is a certification program for low-emitting products and materials commonly used in school buildings, classrooms, and day care facilities. Scientific Certification Systems has developed indoor air quality certifications for building products called Indoor Advantage and Indoor Advantage Gold.

How Green Is Green?

Over the last 25 years, much attention has been given to improving indoor air quality as a result of the practical application of scientific research. Based on studies, papers, and conferences, the new high-performance buildings of today are an embodiment of this work. Architects and engineers are responding with a new consciousness about occupant health, producing new building designs, systems, and specifications. The manufacturing industry is responding with reformulated and new green products. Some independent third-party material certifications are now becoming available to give building material specifiers more confidence in selecting healthy materials, and the construction industry is responding by incorporating green construction methods and adhering to the requirements of the USGBC LEED rating system.

However, much more research is needed to better understand the complex nature of indoor air quality and the human response to specific environments. For example, one area of study that has so far been overlooked is the design of good air quality in healthcare facilities, where the staff spends long hours and where the patients may be more sensitive to the air quality because of their own compromised health conditions. These environments present special challenges based on high-ventilation rates, code-mandated ventilation requirements, and 24-hour operation. It is also important for physicians to be trained to identify and diagnose the health effects of indoor air quality on their patients. Scientific researchers are beginning to establish these connections, and it is now very important to continue to verify this work in practice.

Despite the work yet to be done, there is a major national shift toward green building with significant new knowledge in building-occupant health. With the unfolding of the 21st century, sustainable design and green buildings will become the norm rather than the exception as their design requirements and efficacy are better understood. When this happens, both global ecological health and individual health will have taken an enormous leap forward.

SUSTAINABILITY, HEALTHCARE DESIGN, AND PATIENT OUTCOMES

Judith Heerwagen, Ph.D.

Hospitals are in the business of caring for ill or injured patients and returning them to a more positive state of health. Although there are many institutional and technological factors that influence patient outcomes, it is worth asking how the physical setting and, in particular, sustainable design practices can support patient recovery during hospitalization.

Although there is a growing body of literature on the relationship between the physical hospital setting and patient outcomes, theory and practice has progressed without attention to sustainable design (Ulrich et al, 2004; Rubin et al., 1998). This presentation focuses on how the two fields can be more effectively integrated through application of positive design principles.

A Theory of Positive, Sustainable Design

Positive design integrates risk reduction with experiences that promote emotional, psychological, and social well-being. As noted by Antonovsky in his development of a "salutogenic" approach to health, reducing illness factors does not by itself lead to positive states of health (Antonovsky, 1987). He sees health and illness/disease as lying on a continuum, with different factors contributing to one's location on the continuum. In other words, being healthy and in a state of well-being is not just the absence of risk factors. Health and well-being are supported by a different set of factors.

Although Antonovsky focuses primarily on personal coping factors, his general framework is useful for conceptualizing how the hospital physical environment can be health promoting. First, a brief discussion of well-being is valuable because the concept is not well integrated into sustainable design. The health concerns in sustainable design have centered on a limited number of factors, especially improved indoor air quality. However, there is increasing evidence that a host of building features have positive effects on well-being.

What Is Well-Being?

A recent study of people's perceptions of well-being by Schickler found three key domains:

- Feeling—experiencing positive emotions and sensations, feeling happy and optimistic
- Doing—being actively engaged, moving toward goals, participating in decision making, and experiencing a sense of control
- Being—a state of quiescence, being reflective, or experiencing peace and quietness (Schickler, 2005)

These domains are consistent with the "positive" psychology movement that focuses on the antecedents and consequences of well-being and happiness (Seligman and Csikszentmihalyi, 2000). The positive psychology movement emerged in the 1990s with the growing realization that psychologists knew much about mental and behavioral pathologies, but relatively little about positive behavioral and emotional experience that underlie quality of life and sense of well-being.

The arena of positive psychology is highly relevant to patients, staff, and visitors in healthcare settings. As noted before, the links between health and sustainable design currently focus on improved physical health through improved indoor air quality and reduced exposure to airborne biological or chemical substances. Much less attention is paid to how sustainable design can support positive mental, emotional, and social experiences that underlie concepts of well-being. For design applications, two questions need to be addressed: (1) what experiences underlie a sense of well-being, and (2) what features and attributes of the environment support these experiences?

Experience and Well-Being Needs

Theory and research in biology and behavioral ecology suggests that well-being needs have a strong evolutionary basis (Boyden, 2004; Orians and Heerwagen, 1992) and are linked to specific environmental features and attributes. Well-being needs relevant to the hospital environment include the following:

- Emotional and social support
- Low levels of sensory stimulation similar to those in natural habitats (absent storms or extreme weather)
- An interesting, aesthetically pleasing environment
- Opportunities for recreational activities, including music, dance, and art;
- Connection to nature and natural processes
- Privacy when desired
- Opportunities for rest and psychological recovery

These needs are consistent with surveys of hospital amenities and features that patients want. A recent study by Douglas and Douglas found that patients wanted personal space, a homey atmosphere, a supportive environment, good physical design, access to external areas, and provision of facilities for recreation and leisure (Douglas, 2004). In contrast to the desired environment, many patients experience loss of emotional and social support, boredom, loss of control, absence of natural stimulation, presence of technological stimulation, noise, and feelings of isolation. According to Boyden such environments are in a state of psychosocial deprivation (Boyden, 2004). How can the hospital environment be transformed to create more positive, healthy conditions for patients as well as for staff, and visitors?

Patient Outcomes Linked to Environmental Design

Research in hospital settings shows that a number of environmental features identified in Table AB-1 are associated with improved patient health and well-being outcomes, including improved mood, improved sleep, reduced stress, lower pain levels, and reduced length of stay in the hospital (see extensive reviews of the literature in Rubin et al., 1998; Ulrich, 1991, 1999 ; Ulrich et al., 2004). For instance, key environmental factors influencing health and wellness include the following:

- Sunlight in patient rooms
- Views to sunny spaces outdoors
- Increased individual control over ambient conditions
- Reduced noise with acoustical surface treatments in patient rooms and intensive care units
- Improved privacy and social support with single-bed rooms, more home-like settings for patients and families in hospital rooms, and social spaces that encourage conversation and interaction
- Connection to nature through windows, outdoor gardens, and simulated nature (videos, posters, and paintings)
- Carpeting to soften noise and provide a more comfortable and less slippery surface for walking, especially for elderly or infirm patients
- More pleasing aesthetics and layout, especially a more "homey" or hotel-like room as compared to the institutional look of traditional hospital spaces

Returning to Antonovsky's concept of a health continuum, with disease/illness at one end and health/well-being on the other, it is possible to develop a framework for positive hospital design linked to reduction of health risks and the addition of health and well-being benefits. See Table AB-2.

TABLE AB-1 Well-Being Needs Relevant to Hospital Design and Their Supporting Environmental Features and Attributes

Well-Being Needs	Supporting Environmental Features
Social and emotional support	Visiting and overnight spaces in patient rooms for family and other visitors; social amenities such as cafes, lounges, crafts rooms
Natural levels of sensory stimulation	Reduced noise from people and equipment; more sound absorbing materials such as carpet, acoustical tile, soft surfaces; access to positive sounds (music and nature sounds such as birds, wind, water movement)
	Daylight and access to views of the outdoors
	Operable windows (where feasible) in patient rooms to allow for breezes and connection to outdoor sounds
	Reduced light intensity or use of personal controls in patient rooms
An aesthetically pleasing environment	Reduced look and feel of an institutional setting through better use of color, sensory variation, interesting décor, natural materials, elements of surprise or novelty, and patterned complexity of features
Opportunities for recreational activities	Access to art materials and recreational spaces; music and art therapy; use of headphones to deliver music and reduce unwanted noise; flooring materials that aid movement (e.g., carpeting versus slippery floors)
Connection to nature and natural processes	Views of outdoor nature and sunlight; access to gardens and gardening; indoor sunlight
Privacy	Single occupancy rooms; privacy nooks; visually interesting screening or other moveable partitions at bedside
Rest and recovery	Reduced noise; presence of healing gardens and pleasant outdoor spaces; views of nature from bedside

SOURCE: Heerwagen (2006, unpublished).

TABLE AB-2 The Illness-Health Continuum and Related Environmental Features

Illness and Disease	Health and Well-Being
Poor ventilation	Indoor sunlight
Exposure to chemical	View to outdoor sunlight
toxins and airborne pathogens	Noise-reducing surfaces
Falling accidents	Personal control over ambient conditions
Noise-related stress	Views of outdoor nature from the bed
Noise-related sleep problems	Access to outdoor gardens
Poor maintenance of systems	Social amenities in room and elsewhere
	Reduced institutional look and feel
	Improved overall aesthetics
	Improved support for privacy
	Social spaces that support conversation
	Spaces that support creative activities (art, music)
	Surfaces that improve walking comfort and safety

Source: Heerwagen (2006, unpublished).

What Is Gained by Focusing on Well-Being Needs?

The approach to hospital design suggested in this presentation is not a to-do list. Rather, it suggests a central organizing perspective that begins with theoretical concepts of what it means to be healthy and in a state of well-being. As such, it suggests a system of design interventions and a rationale for their incorporation into hospital design. Hospitals need to do more than avoid harm. As has been rightly pointed out by Ulrich, a hospital needs to provide supportive factors also (Ulrich, 1991, 1999). Whereas Ulrich's supportive design perspective centers on stress reduction and coping, the ideas presented here are derived from theory and research in positive psychology and evolutionary biology. The two perspectives are highly compatible, but look at the hospital situation through different lenses. The value of starting with evolved well-being needs is that the focus is on the patient's experience in a more general way; the process of design then works outward from this central concern. Well-being needs are relevant to all health-care settings. However, the specific needs that should be emphasized are likely to vary, as will the solutions depending upon the context (type of health care facility, patient demographics, culture).

At this time, it is not clear how many of the positive features need to be incorporated into health care buildings in order to move patients and staff to the positive end of the continuum. Do all well-being needs have to be fulfilled or just some? What needs are most important in different contexts or at different ages?

Relevance for Sustainable Design

Many of the design features associated with both risk reduction and health promotion are incorporated into sustainable design approaches, including improved indoor air quality, control of toxic exposures, daylight, and views. However, other factors may run counter to sustainable principles. For instance, noise reduction through use of soft surfaces (carpeting, acoustical tiles, panel systems) may reduce air quality because of increased surface area for particulates to gather as well as a greater need for cleaning. Single-bed rooms and other social amenities, both in patient rooms and elsewhere, may increase the overall footprint of the space and require additional materials. However, if these factors are important to patient recovery, they should be provided in the most sustainable way possible.

THE GREEN GUIDE FOR HEALTH CARE:
A TOOL FOR HIGH-PERFORMANCE HEALING ENVIRONMENTS

Robin Guenther, AIA LEED AP

"If there is one universal truth about hospitals, it is that they are drab, dismal places, not at all designed to soothe and heal" (Alvarez, 2004). This article headlined the health section of the *New York Times* late in 2004 in a cover story that compared the state of U.S. hospital design to hospitals recently constructed in Europe. Specifically, the article showcased the Rikshospitalet in Oslo, Norway, as a "model hospital." Why? What is it about the Oslo structure that yields such a critical comparison to U.S. initiatives? Interestingly, in a 2004 presentation to the American Institute of Architects (AIA) Academy on Architecture for Health, Paul Hyett, RIBA, a UK healthcare architect, summarized the major differences as follows: new technologies in modern architecture produced the high-rise, deep plan, sealed environments that characterize U.S. healthcare buildings, resulting in buildings that are inappropriately low in thermal mass and too heavily dependant on artificial systems.

Architecture is a product of social, economic, political (and yes, environmental) systems and culture. Buildings, ultimately, reflect the goals and values of a society. As Winston Churchill so aptly put it: "We shape our buildings, and then our buildings shape us." Why have our hospitals lost the connection to nature and the vitality that inspires the Rikshospitalet? Does sustainable design have the capacity to transform healthcare architecture? And if so, how do we get there from here?

The Journey to Healing

The annals of Bellevue Hospital tell the history of the development of American medicine. The first public hospital in the United States, Bellevue relocated to the banks of the East River in the early 1800s, away from the dense city, on a site that afforded clean river breezes, fabulous views, and, not incidentally, a convenient river in which to dump the waste. Public health emerged with compelling ideology about the relationship between social conditions in urban environments and disease. By the late 1800s, the city had hired the most famous architects of the day—McKim Mead and White (famous for designing Newport mansions for the wealthy)—to design a major series of pavilion structures to provide health care for the teeming immigrant population of New York City. These pavilions, based on what were known as "Nightingale/Lister principles," featured clean air and fresh water, access to light and views, and focused on allowing nature to heal—with some able assistance from physicians.

By the 1970s, less than 50 years after the completion of the pavilion plan, Bellevue had demolished three of the four pavilions in order to complete the "high-rise hospital," a 22-story hospital building of 1.5 acres per floor (56,000 sq. ft.)—a massive building where fewer than 10 percent of the space had access to natural light, and no space had operable windows. In this generation of health-care construction, the technological advances in medicine and surgery drove larger and larger contiguous floor plates to accommodate the rigors and requirements of the machine. "Systems thinking" (note the diagrams alongside the Bellevue plan) reduced healthcare planning to a series of flow diagrams—flow of people, equipment, and supplies. At the same time, advances in artificial lighting technology and mechanical ventilation supported the redefinition of buildings as "machines for healing." In the span of two generations, we relegated 19th-century ideas about nature and healing, as well as an underlying framework of public health, to "nice if you can achieve it" status while we moved on to the serious work of defining the modern hospital and modern medicine, according to the "machine metaphor."

At the same time, the rapid pace of technological change in health care has outstripped healthcare construction, resulting in facilities that are poorly suited for their function. In urban areas, multiple building campuses spanning a century or more of construction activity are commonplace. The sheer volume of U.S. healthcare construction activity, at more than 100 million sq. ft. annually, attempts to keep the industry current in meeting the growing healthcare needs of citizens. As this happens, the sector consumes an ever-growing segment of the U.S. annual energy bill. As long ago as 1995, healthcare buildings were estimated by the Department of Energy (DOE) to be responsible for 6 percent of total annual energy use (EIA, 1995). Growing concerns about seismic activity on the West Coast has triggered a major healthcare reconstruction program, and the aftermath of Hurricane Katrina will have a huge impact on Gulf Coast healthcare reconstruction.

Health Care and Environmental Health

At the same time that this fascination with technology has consumed the healthcare design world, a quiet transformation has begun in medicine. AIDS, cancer, and the mind-body movement have each challenged mainstream Western medicine's continued stratification into treating "disease" rather than "people." Medicine must increasingly respond to multiple-cause conditions that require multiple therapies. As management of chronic disease replaces the focus on curing the episode, the healthcare industry is entering a period of radical transformation.

The disciplines of public health, environmental medicine, and ecological medicine are emerging as physicians come to understand that our chronic health problems are linked to the environment. Already, we can begin to see the programmatic impacts of these understandings. In design, evidence-based research into the impact of the built environment on therapeutic outcomes and the sustainable design movement are coalescing into a powerful new vision for healthcare architecture.

With regard to planning, evidence-based research surrounding the effect of the built environment on therapeutic outcomes is challenging the prevailing planning models of hospital architecture. In the early 1980s, Edward Wilson's classic book, *Biophelia*, made a strong scientific argument that our affinity for life is the essence of our humanity and binds us to all other living species. This important reconnection informed a wide body of environmental design theory and research into human response to nature, particularly in times of acute stress.

More recently, the work of the Center for Health Design has reinvigorated the discourse concerning linkages between the quality of the built environment and therapeutic outcome. Important evidence-based research by Roger Ulrich linking views of nature to improved recovery rates among cardiac patients has long been recognized in the healthcare planning community. Claire Cooper Marcus' work on the programming and development of healing gardens for healthcare settings is a related area of intense study. Yet despite these important studies, hospitals continue to place more importance on direct horizontal adjacency between operating rooms and recovery spaces than on the therapeutic impact of daylight on patient recovery rates.

As the sustainable design movement generates similar research findings from other building types that promotes closer harmony between buildings and nature—landscape, views, and daylight—evidence-based design and sustainability will coalesce to create a powerful new planning archetype for hospital buildings. The impact of this work on building form can not be underestimated and may provide the impetus for more radical revision of the shape of healthcare buildings far beyond the 20th-century metaphor of "building as machine."

Likewise, public health increasingly links the built environment to health status of the nation's citizens. For example, *Urban Sprawl and Public Health* is concerned with the connections between suburban development and obesity (Jackson and Frumkin, 2004). The media continues to press these linkages: a 2001 cover story of *Business Week* asks the provocative question: "Is Your Office Killing You?" in relation to a myriad of indoor air quality concerns (Conlin, 2000). In fact, building design, materials, and construction practices are responsible for a wide variety of environmental issues that affect human health. Buildings account for somewhere between 40 and 50 percent of fossil fuel emissions (more than the transportation industry, once we reassign transport of building materials), more than 30 percent of raw material extraction, more than 25 percent of potable water usage, and buildings generate more than 30 percent of the solid waste stream. The *Metropolis* magazine cover, in October 2003 proclaimed a truth many of us would rather not be reminded of: "Architects Pollute."

The federal government recognizes these linkages when it defines *green building* as "the practice of increasing the efficiency with which buildings and their sites use energy, water, and materials, and reducing building impacts on human health and the environment through better siting, design, operation, maintenance and removal—the complete building life cycle."

In response to the need for a tool to assist in defining green buildings in the marketplace, the nonprofit U.S. Green Building Council (USGBC), founded in 1993, launched LEED, a third-party certification system for defining and rating sustainable buildings. LEED is not only a point-based metric tool for defining best practices in sustainable design and construction but also a third-party certification system for verifying achievement. Through a rigorous registration, documentation, submission, and certification process, new buildings attain a rating at one of four levels: LEED certified, silver, gold, or platinum. The USGBC sought to attract the top 25 percent of the green buildings in the for-profit commercial building marketplace; almost immediately, however, state and municipal governments began adopting the rating system as a minimum standard of construction (Progress Report on Sustainability, 2003). LEED is not explicitly health focused.

Although the LEED program has been successful in registering and certifying commercial office buildings, university buildings, and the like, its adoption by health care has been relatively limited. In December 2003, Boulder Community Foothills Hospital became the first healthcare facility to achieve LEED Silver certification. This 60-bed, $53 million, 200,000-sq.-ft. facility was followed less than one year later by the Discovery Health Center (LEED certified at 28,300 sq. ft.). In February 2005, the Emory Winship Cancer Center (LEED certified with 260,000 sq. ft., costing $75.7 million) became the third healthcare facility LEED certified. Each manifests core principles of using healthy and low-emitting materials, abundant daylight, efficient energy systems, and careful regard to site considerations, among other green design features. At this writing, there are a total of 57 healthcare projects registered in the program, representing just over 17 million sq. ft.

The reasons for this, we believe, lay in three important areas:

• There was no explicit connection to human health; green building was viewed as being "good for the environment"—no explicit value connection.

• Substantive differences exist between commercial office buildings and acute care facilities that the LEED rating tool did not recognize—a poor fit.

• Healthcare construction is accomplished by a specialized segment of the architecture and design field that was relatively disconnected from the environmental building movement—lack of education.

The *Green Guide for Health Care* was developed to address the challenge of moving the green building agenda into health care. The charge: to make an explicit "health value" connection through a tool that "fit" the sector, and educate the industry to use the tool.

The Environmental Footprint of the Healthcare Industry

To develop a tool to fit the sector, it was necessary to understand the environmental footprint of the healthcare industry. So, what is the environmental footprint of the healthcare industry? How is it measured?

In 1996, medical waste incineration was named as the second leading source of dioxin emissions in North America (EPA, 1996). Since then, the number of medical waste incinerators has decreased from 5,600 to just over 100, as the industry has moved to clean up its operation. The use of mercury in the healthcare sector has sharply declined since the late 1990s, with a voluntary goal of virtual mercury elimination set for 2005. In operations, the industry has achieved compelling victories. As a result of that work, we knew that many hospitals had "environmental champions" that were making great progress in pollution prevention initiatives.

In October 2000, Health Care Without Harm, Kaiser Permanente, and Catholic Healthcare West came together at a landmark conference, "Setting Healthcare's Environmental Agenda" (SHEA), in Oakland, California. The conference began with a challenge issued by Michael Lerner, Ph.D.: "The question is whether healthcare professionals can begin to recognize the environmental consequences of our operations and put our own house in order. This is no trivial question" (Lerner, 2000).

David Lawrence, chairman and CEO of Kaiser Foundation Health Plan, the largest nonprofit healthcare system in the United States, said, "Just as we have responsibility for providing quality patient care [and] . . . keeping our facilities and technology up to date, we have a responsibility for providing leadership in the environment" (Lawrence, 2000). Lloyd Dean, president and CEO of Catholic Healthcare West, agreed, adding this challenge: "We will not have healthy individuals, healthy families, and healthy communities if we do not have clean air, clean water, and healthy soil" (Dean, 2000).

As pollution prevention initiatives took hold, explicit connections between health care's environmental footprint and human health were being increasingly recognized. *Greener Hospitals*, published in Germany by Bristol Myers Squibb, included the triangular diagram linking environmental impacts of healthcare delivery to increased environmental stress and illness.

It is imperative for the design of "high-performance healing environments" that the alignment between sustainability and health care's primary mission—to heal—is explicit. In 2002, the American Society for Healthcare Engineering (ASHE) issued a green healthcare construction guidance statement that made just such an explicit statement. The introductory statement of principles asserts the importance of protecting health at three scales:

- Protecting the immediate health of building occupants
- Protecting the health of the surrounding community
- Protecting the health of the larger global community and its natural resources

The Green Guide for Health Care

How will we measure our performance in the area of environmental stewardship? How will we objectively assess the size of our environmental footprint? The *Green Guide for Health Care* was developed to respond to these questions. It is the first healthcare-specific metric self-certification tool for guiding continuous environmental improvement in the healthcare construction and operations world.

Conceived of as a tool for organizations and their project teams to use in designing and operating facilities, the *Green Guide* emphasizes environmental and public health issues in evolving strategies for creating "high-performance healing environments." It synthesizes program, planning, and materials/systems strategies, as outlined above, to provide the most comprehensive guide to the design of healing environments yet undertaken by this industry.

The *Green Guide* adopted ASHE green healthcare construction guidance statement of principles. It reaffirmed a principle of precaution, echoed in medicine and international sustainable design policy. From LEED, with permission, it gained credit structure, content, and organization. Building upon the work of Hospitals for a Healthy Environment (H2E), it reinforced the commitment to the EPA memorandum of understanding and defined a comprehensive approach to healthcare operations. From the early healthcare green building adopters, it evolved rigorous materials evaluation requirements, particularly with regard to emissions and persistent, bioaccumulative, and toxic (PBT) compound avoidance. It reinforced principles of evidence-based design, through emphasis on daylighting, acoustics, and places of respite.

Finally, the *Green Guide* recognizes that the healthcare industry is in the early stages of sustainability development and is adverse to regulatory enactment.

The *Green Guide* does not establish minimum achievement thresholds. It is a self-certification tool with an emphasis on promoting best practices within the industry by instilling a culture of internal assessment, evaluation, and continuous improvement. Since its initial release in 2003, the *Green Guide*'s goal of transforming the healthcare sector's building portfolio into healthy, high-performance healing environments is being realized through a measured approach grounded on best practices, industry partnerships, and implementation feedback. As an evolving document, it continues to be refined through an update process.

What We Are Learning from the *Green Guide* Process

First, the healthcare industry is largely uninformed about the explicit linkages between sustainable building strategies and human health. Pilot participants and registrants consistently report that the health issue and resource information contained in the *Green Guide* is important new information for many of them. More education within the medical, nursing, and healthcare executive communities is required to continue to complete the linkage between sustainability, health, and mission.

Second, we are learning that sustainable building measures are happening in the industry outside of the LEED tool, particularly projects with this explicit health mission. However, we need to develop educational tools and market incentives to move hospitals and healthcare organizations from tier 2 to tier 3, as defined below:

- Tier 1—Minimum local, state, and national environmental regulatory compliance
- Tier 2—Beyond compliance to measures that save money
- Tier 3—Informed by the inextricable link between environment and human health and moving beyond both compliance and monetary savings with a long-term plan to reduce environmental footprint—a "triple bottom line" approach (Schettler, 2001)

Third, that as a market transformation tool, the *Green Guide* is pivotal at informing product development. Many of the *Green Guide* registrants represent product manufacturers eager to be proactive in producing healthier, sustainable products for a range of applications. The number of product innovations such as PVC avoidance, brominated flame retardants (BFR) elimination, and sustainably forested and recycled or rapidly renewable products are increasing.

Finally, we are learning that project success may be related to the explicit understanding and development of an approach that prioritizes health and integrates design and operation into an environmental health mission statement. LEED implies that an "integrated design" process yields greater project success; the *Green Guide* requires the development of a mission statement and rewards an "integrated design process." More explicit research and guidance tools that link design and operations for a reduced environmental footprint is necessary to empower teams to embrace both aspects of building performance.

What Is the Challenge?

We must begin, as an industry, by demanding life cycle solutions from the manufacturers and industries that service and support us. We must vote with our specifying and purchasing power (which, incidentally, is significant), and support approaches that "solve rather than alleviate the problems that industry makes" (McDonough, 1998). The initiatives that Kaiser Permanente has shown in using the power of their contract dollars to drive significant environmental improvements for building products is a shining example of the power of our collective voice for healthier, sustainable building materials. We must not be forced to resolve insoluble statistical conflicts in which materials are competing on the basis of being the lesser of many evils rather than on the basis of being good.

Secondly, we must support the development of clear, universal materials assessment methodologies that take into account the "hidden" health costs associated with toxic materials and industrial processes. Life cycle assessment methodologies, as currently developed, can offer objective measurement of raw materials and energy flows, but are silent or scientifically inadequate regarding inclusion of environmental health and toxicity issues. The healthcare industry must assist in evolving life cycle tools that appropriately include full health costs.

Finally, the industry must seek to evolve cost models that recognize health costs as an important component of the price we pay for our buildings today; these models must also show health benefits as creating value. In so doing, the industry can author a powerful tool for the wider real estate industry and support market transformation at the forefront of innovation and building materials development.

REVIEW OF GREEN AND HEALTHY SCHOOLS: COSTS, BENEFITS, AND IMPLICATIONS

Gregory Kats

Some 50 million students spend their days in schools that are too often unhealthy and that restrict their ability to learn. A recent and rapidly growing trend is to design schools with the specific intent of providing healthy, comfortable, and productive learning environments. These green, high-performance schools generally cost more to build, which has often been considered a major obstacle at a time of limited school budgets and an expanding student population.

A December 2006 national review of 30 green schools and an analysis of available research demonstrate that green schools cost 1.5 to 2.5 percent more than conventional schools, but they provide financial benefits that are 20 times as large (Kats, 2006). These financial benefits include energy and water savings, reduction in costs associated with waste and emissions, increased student learning and future earning, reduced incidence of student asthma and other illnesses, reduced costs of teacher turnover, and net employment gains for the state. Most of these benefits relate to improved health, enhanced student learning, test scores and earnings, and reduced teacher turnover.

Conventional schools are typically designed to just meet the building codes, which are often incomplete. The design of schools to meet minimum code performance tends to minimize initial capital costs but delivers schools that are not designed specifically to provide comfortable, productive, and healthy work environments for students and faculty.

Few states regulate indoor air quality in schools or provide for minimum ventilation standards. A chronic shortage of funds in schools means that schools typically suffer from inadequate maintenance, resulting in degradation of basic services such as ventilation and lighting systems. Not surprisingly, a large number of studies have found that nationally, schools are unhealthy, which increases illness and absenteeism and brings down test scores. Green school design provides an extraordinarily cost-effective way to enhance student learning, reduce health costs and, ultimately, increase school quality and competitiveness at both the student and state level.

The main reason for cities and states to adopt green building requirements is to cut costs, improve services, and address a broad array of challenges, such as

- the high and rising cost of energy,
- worsening power grid constraints and power quality problems,
- increasing cost of waste, water, and waste disposal and associated costs of water pollution,
- continuing state and federal pressure to cut air pollution,
- rising concern about global warming,
- reversing the alarming rise of asthma and allergies in children, and
- increasing state competitiveness in quality-of-life indicators such as air and water quality, quality of schools, and the skills of its workforce.

This analysis finds that green schools provides an extremely cost-effective way to help address all these challenges. The financial benefits of green schools are 10 to 20 times as large as the cost. Green school construction costs 1.5 to 2.5 percent more than conventional school construction, almost $4 more per sq. ft. for a typical $25 million, 125,000 sq. ft. school built for 900 students. The financial savings are about $70 per sq. ft., 20 times as high as the cost of going green (Table AB-3) (Kats, 2006). Only a portion of these savings accrue directly to the school. Lower energy and water costs, improved teacher retention, and lowered health costs save green schools directly about $15/sq. ft., about four times the

TABLE AB-3 The Financial Benefits of Green
School Design ($/sq. ft.)

Energy	$14
Emissions	$1
Water and wastewater	$1
Increased earnings	$37
Asthma reduction	$4
Cold and flu reduction	$4
Teacher retention	$4
Employment impact	$3
Total	$68
Costs of Green Design	$4
Net Financial Benefits	$60–$70

SOURCE: Kats et al. (2006).

additional cost of going green. Financial savings statewide are significantly larger, and include lower energy costs, reduced cost of public infrastructure, lower air and water pollution, and a more skilled and better compensated workforce. The majority of these savings results from improved health, comfort, and learning performance in green schools.

Green schools provide a range of additional benefits that were not quantified in this report, including reduced teacher sick days, reduced operations and maintenance costs, reduced insured and uninsured risks, improved power quality and reliability, increased state competitiveness, reduced social inequity, and educational enrichment. There is insufficient data to quantify these additional benefits, but they are significant and, if calculated, would substantially increase the recognized financial benefits of greening schools.

Despite limits in data and need for additional research, there is now very substantial experience with high-performance schools in Massachusetts and nationally. A large body of documented studies and experience allows quantification of costs and benefits of green schools. For example, there are over 1,000 studies that examine the impact of high-performance design features such as better lighting, temperature control, and improved indoor air quality on health or productivity. Analysis of the costs and benefits of 30 green schools nationally, and use of conservative and prudent financial assumptions in analyzing available data, provides a clear and compelling case that green schools today are extremely cost-effective from a financial standpoint. Largely because of improvements in health, attendance, test scores and learning environment, building green schools is today significantly more fiscally prudent and lower risk than building conventional unhealthy, inefficient schools.

IMPLEMENTING THE DECISION TO BUILD GREEN IN UNIVERSITY MEDICAL CENTERS

Roger A. Oxendale

In the late 1940s, Frank Lloyd Wright offered an abrupt, two-word answer when asked how he would improve Pittsburgh. He said: "Abandon it." Fortunately, our city leaders took that excessive assessment as the metaphor I believe it was—not to abandon an entire city, but to rid itself of old ways of thinking; that is, to create a new environment that would clear our region of the industrial waste, heavy smoke, and soot created by the massive steel mills that were turning Pittsburgh into an industrial wasteland of putrid air and contaminated waterways.

History bears out the foresight of Pittsburgh's government and civic leaders who moved to sustain the long-term health of a people and the region in which they live. They did that with bold strokes, such as requiring businesses and homeowners to switch from using pollution-causing coal to gas or other smokeless fuels for heating.

Those were great strides that enabled Pittsburgh to move forward with courage and conviction—all the while creating an extraordinarily livable, diversely economic region, which now has one of the most highly regarded, sophisticated healthcare systems in the world.

For this progress to continue as we move ahead in this rapidly changing century, we have a responsibility to improve the health of our region, a responsibility we took seriously as we began making plans for construction of the new Children's Hospital of Pittsburgh as part of the University of Pittsburgh Medical Center (UPMC).

Our vision was to create a model that would lead to the transformation of health care in areas of chemical and hazardous waste management; air quality; green construction and retrofits; renewable energy; energy and water conservation; and housekeeping (Dick Corporation, 2006).

From the early stages of planning we realized that few, if anyone else, in health care were thinking as broadly as we were about what it means to be green. Yes, there are hospitals in the country that incorporate significant environmentally sustainable improvements using green materials; optimizing energy performance; and reducing, reusing, and recycling chemicals and supplies.

We decided to go well beyond that because our commitment to health care stretches from our hospitals' walls into the community and region in which our patients and families live.

At Children's Hospital, we held a series of internal forums on pediatric environmental health to begin to look at how we could provide our community with comprehensive environmental health education through community projects and programs.

We will do this by linking our extraordinary scientific research to advanced treatments for our children; training our residents and providing continuing medical education to our physicians to incorporate green practices and treatment for the improved health of our kids, their homes, and communities; and educating—within family units, in the schools, in communities—so that that the UPMC healthcare system is integrated into how we think and live in western Pennsylvania.

Thanks to a commitment to green by the Pittsburgh-based Heinz Endowments—our partner in building a national model of children's environmental health and hospital sustainability—we at Children's Hospital were afforded the opportunity to dream big and then to develop a careful plan for our leadership. Our vision also was endorsed by Pennsylvania's elected officials when Governor Ed Rendell awarded Children's Hospital $5 million toward construction of a green pediatric hospital.

So, we proceeded, embracing the knowledge offered by other like-minded institutions, including the University of Pittsburgh's School of Engineering, Sustainable Pittsburgh, the Pittsburgh Green Building Alliance, and the University of Pittsburgh Cancer Institute's Center for Environmental Oncology.

We are the first pediatric hospital in the country to register for LEED certification. By doing that, we are helping to establish a LEED guide for healthcare institutions. Moreover, we are creating a healthcare model for environmentally sustainable engineering in a city that is recognized as a leader in the worldwide green building movement.

Pittsburgh is home to the world's first and largest certified green convention center—the David L. Lawrence Convention Center. Not only that, Pittsburgh now has 40 buildings that are either LEED certified or registered.

Our challenge, as we strive for silver and gold LEED certification, is for Children's Hospital to incorporate construction of a new, technologically advanced research building, the retrofitting of a partially demolished building for an equally high-tech clinical facility, as well as the renovation and retrofitting of older buildings throughout the UPMC system.

At the same time, we know we can not work in a vacuum; that is, we can not expect to move into our new Children's Hospital of Pittsburgh in late 2008, the expected date of completion, without having already incorporated environmentally sustainable practices at our existing facilities.

For that, we hired a team of researchers at the Center for Building Performance and Diagnostics at Carnegie Mellon University to investigate greening opportunities for the operation and maintenance of Children's Hospital and other UPMC hospitals today.

This has provided us with an opportunity to establish benchmarks for environmental sustainability as well as identify benchmarks for innovation as we plan for the future. These are significant measures that will enable us, through the University of Pittsburgh's School of Engineering, to develop baseline data so we ultimately will be able to track the environmental impact of the new Children's Hospital—one that is a highly efficient, sustainable facility that incorporates water and energy conservation, improved indoor air quality, and green building materials and cleaning practices.

In preparation for LEED certification, we are proceeding with key environmental considerations at our new hospital. And, given the work being done at the University of Pittsburgh Cancer Institute and the Graduate School of Public Health, we are moving ahead in building relationships in multiple areas, including an environmentally preferable purchasing policy; reductions in the use of toxic chemicals, the consumption of water and energy, and the volume of all waste streams; improvements in indoor air quality and work atmosphere; and coordination of efforts to reduce all forms of pollution, including vehicle emissions.

The measures that Pittsburgh's leaders took almost 60 years ago to clean up our region enables us throughout UPMC to promise a future where hospitals will be models of disease prevention and cure.

And, we begin with construction of the new Children's Hospital of Pittsburgh—where the greening of this pediatric institution will enable us to continue to transform the lives of our young patients so they can return home to a healthy future.

THE COMPELLING BUSINESS CASE
FOR BETTER HOSPITAL BUILDING

Derek Parker FAIA, RIBA, FACHA

Building a new facility is usually the biggest capital investment a chief executive officer, medical staff, and board of trustees will ever make. Hospitals will spend more than $12 billion this year on new construction and, by 2010, spending on new hospital construction is expected to increase to $16 billion to $20 billion annually (Sadler, 2004). With so much at stake, the time is right for hospital leaders to spend a little more time and money to not just build a new hospital, but a better hospital—one that will actually save significant dollars in the long run. It costs a lot of time and money to build a poor hospital. It costs only a little more time and money to build a hospital that contributes to clinical, financial, and satisfaction outcomes based on evidence-based design.

But, in the current health care economic environment with capital so difficult to obtain, you might ask: Are these good ideas affordable? Is there a business case for building better hospitals? The answer is yes. Based on published evidence and the experience of pioneering organizations using evidence-based design to construct new facilities, we have analyzed the data and designed a hypothetical "Fable Hospital."

Getting Started

A healthcare executive or trustee who wishes to follow a path similar to Fable Hospital might ask, "How best to begin?" It begins with the vision that positive effects on patients, staff, and the community will occur through a collaborative commitment to combining the best design evidence with the core values and belief systems of the organization. Thus a first step is to formally define and widely disseminate this vision and keep it in front of organizational members at all times.

The next step is to become familiar with the work of the pathfinders who are blazing the trail for others. This can include reading, attending conferences, and taking benchmarking tours of exemplary projects. One wise measure would be to assure that the organization's guiding coalition grasps the importance of an evidence-based course for decision making on design and construction projects. Another would be to assemble a strong collaborative team of advisors who have the complementary skills and experience to rigorously follow such a course. A team of programming consultants, architects, engineers, and interior designers who value evidence-based design might be bolstered with social scientists, such as an environmental psychologist or an expert in performance improvement. The prudent executive should be prepared to invest extra time preparing a sophisticated description of the project that goes beyond a simple listing of proposed space requirements. It is helpful to be able to describe a project's goals and objectives with clarity, including hypotheses concerning outcomes expected from the design.

Resistance to a process that differs from prevailing practice can come from almost any source. In addition to the predictable resistance to any form of change, the team can expect to be challenged at first by skeptics who will question the evidence, the financial assumptions, and the link between facility design and clinical outcomes. This is why a certain amount of study and a team accustomed to rigor will be useful. The challenge to financial assumptions will require careful analysis and cautious budgeting that avoids overreliance on previous budget or cost models. It would be wise to involve the external consultants early in the process to gain the maximum benefit from their experience.

A typical barrier to success is expecting a project to neatly fit into the same budget and schedule as a conventional project, when in fact it likely will require an extended predesign phase to properly define the scope; identify, analyze, prioritize, and integrate design innovations; and plan an assessment protocol. The team should be prepared to do more sophisticated life cycle costing than occurs in a conventional project, as fewer decisions will be based exclusively on the lowest first cost. A savvy executive will insist on using multiple before-and-after measures to assess the project, including financial, clinical, and satisfaction indicators.

BUILDING-RELATED HEALTH EFFECTS: WHAT DO WE KNOW?

Ted Schettler

Hospital buildings provide space for health care, employment, residence, shelter, and comfort. Building design, construction, operations, and maintenance influence the indoor environment and the health and well-being of staff, patients, visitors, and other occupants.

Design and construction decisions also affect the environment and public health regionally and even globally. Materials extraction, product manufacturing, transportation, use, recycling, and disposal influence air and water quality, land use, and can contribute to ozone depletion and climate change. The health of workers in the supply, production, and disposal/recycling chain, as well in building construction, operations, and maintenance, is also affected.

This paper primarily addresses the influence of buildings on the health of occupants. It briefly touches on more far-reaching concerns, including the appropriateness of certain activities related to health care.

The Indoor Environment

Building-related comfort and health are directly related to indoor environmental quality, which is determined by combinations of temperature, temperature gradients, humidity, light, noise, odors, chemical pollutants, personal health, job or activity requirements in the building, and psychosocial factors. That is, buildings are complex dynamic systems of multiple interacting factors that determine the state of the system at any given time.

Microenvironments within buildings may be highly relevant determinants of health impacts among occupants. Spatial heterogeneity among a mixture of relevant variables makes it difficult to study and understand causal health-related relationships (Spengler and Chen, 2000).

Much work on building-related health focuses on combinations of temperature, humidity, ventilation, and indoor air pollution. Air pollutants include volatile organic compounds (VOCs); semivolatile organic compounds (SVOCs); microbial VOCs (MVOCs); particulates; nitrogen oxides; ozone; carbon dioxide; and biological agents such as bacteria, viruses, and fungal spores. Many air pollutants are generated indoors, and others infiltrate from the outdoors. These factors interact in multiple combinations that vary over time and place, even within the same room or building, making it difficult to understand the extent to which each contributes to health outcomes.

For example, assessments of exposure to indoor air pollutants that assume homogeneous concentrations in a room will miss important concentration gradients around point sources of emissions. Concentrations may vary by several-fold, depending on proximity to an emitting source (Furtaw et al., 1996).

Building design, operations, and maintenance must be considered collectively. Design and construction choices will influence operations and maintenance in ways that make building-related complaints more or less likely.

Many studies that attempt to examine building-related illness are limited by their design (e.g., cross-sectional surveys are common and are limited by several kinds of bias), lack of quantitative exposure information, subjectivity in outcome measures, and uncertainty about what potentially causal factors should be measured. Further, because of interactions among multiple building related factors, commonly used statistical techniques do not lend themselves to the analysis. Models based on principal component analysis or structural equation modeling show some promise, but will need further work before being generally applicable (Pommer et al., 2004).

Building-Related Illness, Building-Related Symptoms, Sick-Building Syndrome, and Multiple Chemical Sensitivity

Sharp distinctions between health and comfort are not readily apparent and may not be appropriate. Building-related illnesses include specific diseases such as Legionnaire's disease, which can be traced to a single source or cause. Building-related symptoms include (EPA, 2006)

- mucous membrane symptoms (blocked or stuffy nose, dryness of the throat, rhinitis, sneezing, dry eyes),
  - headache, confusion, difficulty thinking and concentrating, and fatigue;
  - cough, wheeze, asthma, and frequent respiratory infections; and
  - allergic reactions, such as dry skin.

The term *sick building syndrome* (SBS) is used to describe situations in which building occupants experience acute health and comfort symptoms that appear to be linked to time spent in a building, but often no specific cause can be identified. Complaints may be localized in a particular zone or widespread throughout the building. SBS is sufficiently common and has been sufficiently described to have attained robust stature in medical and architectural disciplines.

To further complicate analyses, some people seem to be particularly sensitive to a wide variety of environmental contaminants at relatively low concentrations. In some of these people, a diagnosis of multiple chemical sensitivity (MCS) suggests that it is virtually impossible to separate assessments of the quality of the indoor environment from the unique vulnerability of some building occupants. The pathophysiology of MCS is uncertain and controversial, although an increasingly robust scientific database supports the importance of this phenomenon (National Research Council, 2002). It is, therefore, difficult to draw a distinct line between a building with an unhealthy indoor environment and one in which a subset of building occupants appear to have heightened sensitivity to often poorly defined but ordinary environmental contaminant levels.

Building-Determinants of Indoor Environmental Quality, Comfort, and Health

Building Material Emissions and Reactivity

Building operating conditions and products used in building design and operation create an environment in which complex emissions and chemical reactions can occur. Direct emissions from building materials (primary emissions) are generally highest soon after manufacture and construction and diminish thereafter. Secondary emissions are caused by the actions of other substances or activities on the material. For example, moisture, alkali in concrete, ozone from electronic equipment, or cleaning materials can influence emissions from building materials. Secondary emissions may be a chronic problem (Sundell, 1999).

Cooler surfaces on a wall can increase local relative humidity facilitating emissions from wall-covering material. Humidity or dampness in concrete floor construction facilitates alkaline degradation of di-ethyl-hexyl phthalate (DEHP), a plasticizer used in polyvinyl chloride (PVC) floor covering as well as other PVC products.

Ozone that gains entrance from the outdoors or that is emitted from photocopiers or laser printers can react with unsaturated double bonds in various polymers to create aldehydes and ketones. These secondary emissions may be highly reactive, and irritate skin and mucous membranes of building occupants (Wolkoff et al., 1997).

Nitrogen oxides from outdoors or generated from photocopiers or laser printers can also react with a variety of VOCs to form irritant compounds, including aldehydes (Wolkoff et al., 1997). Highly reactive free radicals are also formed by reactions of NO_2 and ozone with unsaturated compounds. Many of these compounds are not easily measured, yet they may be highly relevant in terms of health effects.

Indoor Pollutants Associated with Building Operations and Maintenance

Building design decisions can also influence which products are used in routine building operations and maintenance, and thus influence indoor environmental quality. Some cleaning products contain respiratory tract sensitizers or irritants. Even cleaning products promoted as "greener" sometimes contain citrus or pine-based materials that can themselves, or in reaction with oxidants such as ozone, contribute to indoor air pollution. Occupants of buildings cleaned more often that once weekly tend to report fewer building-related symptoms (Skyberg et al., 2003).

Building and landscape design can influence the likelihood of indoor pest problems. Routine use of integrated pest management strategies can reduce indoor and outdoor pesticide use, thereby contributing to improved indoor environmental quality.

Ventilation

High- or low-ventilation rates can have a significant impact on symptoms. Limited evidence suggests that ventilation rate increases up to 10 L/s per person may be effective in reducing symptom prevalence and occupant dissatisfaction with air quality; higher ventilation rates are not effective (Spengler and Chen, 2000). But because of complex relationships among ventilation rates, contaminant levels, and building-related health complaints or satisfaction with air quality, the use of ventilation as a mitigation measure for air quality problems should be tempered with an understanding of its limits.

Dampness and Humidity

Building dampness can facilitate mold growth, particularly on surfaces with organic material that can serve as a nutrient source. MVOCs can also be emitted from heating, ventilation, and air-conditioning (HVAC) systems. Fung and Hughson reviewed all English language studies (n = 28) on indoor mold exposure and human health effects published from 1966 to 2002. They concluded that excessive moisture promotes mold growth and is associated with increased prevalence of symptoms due to irritation, allergy, and infection. However, methods for assessing exposure and health effects are not well standardized (Fung, 2003).

Surface Materials

Several studies show a correlation between certain materials on interior building surfaces and risks of asthma, wheezing, or allergy. Materials that may be causally related to these symptoms include PVC flooring and wall coverings, new linoleum, synthetic carpeting, and particle board (Jaakkola et al., 2004). Increased risk of childhood risk of bronchial obstruction, wheezing, and allergic symptoms is reported associated with PVC plastic and plasticizer-containing surfaces. (Bornehag et al., 2004a; Jaakkola et al., 1999; Norbäck et al., 2000; Oie et al., 1999; Tuomainen et al., 2004).

Particulate Air Pollution

Particulate indoor air pollution is of variable size and composition. Particulates may contribute to building-related symptoms in occupants, but the relative contributions of particle size, particle mass, and particle composition are uncertain (Christensson et al., 2002). High-speed floor polishing can contribute significantly to airborne particulates, depending on the equipment used and the nature of the surface material (Bjorseth et al., 2002; Roshanaei and Braaten, 1996).

Health Impacts Beyond the Building

It is also important to acknowledge that hospital design, construction, and operating decisions can have far-reaching public environmental health effects from water and energy consumption, materials transportation, and occupational health concerns throughout the materials supply chain.

Releases of environmental pollutants from materials extraction, manufacturing, and disposal practices can have regional and even global consequences for public environmental health. Building designers have an opportunity to influence worker and public environmental health through informed materials selection and attention to worker and social justice concerns.

In addition, it is essential to begin to address explicitly the long-term public and environmental health impacts of healthcare activities themselves. Those activities are rarely subject to the same scrutiny to which we subject the building infrastructure.

In the United States, expenses related to health care make up about 15 percent of the gross national product. This amount is growing annually, and much of the growth can be attributed to the development of new technologies, each with its own implications for public environmental health.

Resource extraction, materials manufacture, and disposal are responsible for most human impacts on the natural world. The scale of healthcare activities and life cycle impacts of related flows of materials contribute substantially to environmental degradation. High-tech equipment, pharmaceuticals, transportation, and water and electricity consumption in health care have major environmental impacts. Despite the commitment of most countries to growth, material throughput must be drastically scaled back in order to achieve sustainability. The healthcare system must do its share.

Pierce and Jameton have made a strong argument for health care's particular ethical responsibility (Pierce and Jameton, 2004). Marginal improvements in materials policies may help, but a fundamental reexamination of the scope of clinical services is also required. This may inevitably lead to concerns about rationing, but rationing, according to Pierce and Jameton, should not be thought of as less than optimal care but rather as sustainable optimal care, if the healthcare industry is going to meet its ecological responsibilities.

Conclusions

Buildings are complex dynamic systems composed of multiple materials assembled and operated in ways that create an indoor environment with considerable heterogeneity in space and time. Building-related illnesses result from multiple factors that are often difficult to quantify and that interact in complex ways. Considerable additional research is necessary in order to advance the understanding of building-related health effects. Statistical techniques used in the analysis of complex dynamic systems may be helpful and should be further explored.

Although it is difficult to establish clear-cut evidence-based guidelines for all aspects of building design, construction, and operation, several themes emerge from the published literature. Low-emitting materials should be selected. Materials that might support mold growth should be reduced. Building design, construction, and operations should ensure that moisture does not accumulate. Material selection should be influenced by cleaning requirements and the extent to which cleaning may contribute to VOC and particulate concentrations. Low-emission materials, along with appropriate ventilation, temperature and humidity control, will contribute to improved indoor air quality.

Individual, community, and ecological health are interpenetrating. They are influenced by building design, construction, and operating decisions and should be routinely assessed during planning stages. Along with attention to direct and indirect impacts of building design, construction, and operating decisions, a fundamental reexamination of the scope of clinical services is also required, if the healthcare industry is going to meet its ecological responsibilities.

BUILDING GREEN ON A LARGE SCALE

Scott Slotterback

Often culture drives decision making. Typically we get answers to only the questions we ask. At Kaiser Permanente we believe it is time to start asking different questions. It is time to imagine a future filled with potential and ask the questions that will help us realize that vision. We plan on being a part of that positive future. As Marshall McLuhan said, if we drove the way we typically plan we would spend most of our time looking into our rearview mirrors and we would all crash our cars. All too frequently when we plan the future, we focus on the past, so we can build on a strong foundation, correct our prior mistakes, and gradually make transitions. In slower times this was quite effective. However, with today's rapid pace of change, we need to look into the future just to stay current. This is especially true when building green on a large scale. As we set out to design and construct buildings that embody Kaiser Permanente's vision for environmental performance, we seek answers to questions the marketplace has not been asking. We ask for products that do not yet exist. We create incentives for manufacturers to provide these products. And we buy the products that meet our grueling criteria. Building on a large scale does have its advantages, and we are using these advantages to facilitate a market transformation in green buildings for health care.

How big is the "large scale" I am talking about? I must admit being in Washington, DC, where people commonly talk about trillions of dollars being spent, it is a little intimidating to talk "large scale." I am not talking about trillions, but I am talking about billions and millions. Kaiser Permanente plans to spend more than $20 billion on its capitol program over the next 10 years. We currently have

8.3 million members in nine states and the District of Columbia. We have 60 million sq. ft. of occupied space in over 900 buildings. We are planning to build 14 seismic replacement hospitals, six new hospitals, three major hospital bed expansion projects, and numerous hospital renovation projects and new medical office buildings along with the central utility plants and the parking structures needed to support them. So to me, that seems to be reasonably large scale.

Because six million of the of the eight million members of Kaiser Permanente reside in California, my colleagues here in the mid-Atlantic region often point out that we are somewhat less of a household name here in Washington than we are in my home town of San Francisco, California. So I will give you a quick overview of how we are structured, since even though we are large, we may not be familiar to you. The organization known commonly as Kaiser Permanente is actually three companies in one. Kaiser Foundation Health Plan, Inc. is a nonprofit insurance company. Kaiser Foundations Hospitals is also a nonprofit company that manages the hospitals, and the Permanente Medical Groups are the for-profit associations of physicians. Together, these three organizations make up our integrated model of care; from insurance carrier, to physician, to hospital and staff. So those of us, like me, who are focusing on the design and construction of the hospitals and other buildings are very closely tied to the users of these buildings: our physicians, staff, and members. As a result we care a great deal about their health and safety.

Building green on a large scale is not only about the health and safety of our physicians, staff, and members, but it is also about the health of our communities. To lead this effort Kaiser Permanente established an Environmental Stewardship Council, which is charged with achieving Kaiser Permanente's vision for environmental performance. Our vision is stated in one far-reaching sentence: We aspire to provide healthcare services in a manner that protects and enhances the environment and health of communities now and for future generations.

For us green building is not limited to impacts our buildings have on the people who use them. Green building also includes the downstream impacts on the communities that make the building materials and our community at large.

How do we define green? Actually, we have turned to others to help us clarify that concept. In 2002 we used the ASHE Green Guidance statement as the foundation for our own *Eco Toolkit*, a document that links Kaiser Permanente's robust design standards program to the green practices identified in the ASHE statement. Today we are using the *Green Guide for Healthcare* (GGHC) as a green training tool, a success-measuring tool, and as the foundation for our next generation of our *Eco Toolkit*. The GGHC provides us with a national standard to objectively measure our success.

What are we doing to implement our grand vision of a positive future? Let me give you a few examples.

I would like to talk to you about the numerous green initiatives Kaiser Permanente is implementing on building projects, but there is not time in this presentation, so I will summarize a few of them and discuss one or two in more detail to illustrate how we are overcoming institutional barriers to building green.

Before I focus on specific examples, I would like to discuss how we are dispelling one of the institutional barriers to sustainable design—increased costs. How many of you have been told at one point or another that building green costs more? Does this have to be true? We do not think it does, and we are proving that many green measures can be implemented without adding costs. Many of the green measures we are testing on projects that we are building today are cost neutral, and some are significantly reducing costs.

Here is a quick summary of the results of some of these efforts. Since 1998 we have had an alliance program that brings together architects, engineers, and contractors to work with our physicians, staff, and other owner representatives, starting in the early phases of the project to provide an integrated design process. Having all these stakeholders working together builds a shared understanding of the value of green measures and enhances their continued implementation when the building is completed and occupied.

Permeable paving, which allows water to filter back into the aquifer, is currently being tested on a 50-acre new medical campus. Although the paving is more expensive than conventional paving, when we looked at the issue systemically we found that using permeable paving eliminated the need to connect the project to the city's storm water drainage system. This saved us the cost of running almost a quarter mile of storm water piping which saved us a significant amount of money.

Our design standards recommend that drought tolerant native species be used in landscaping to reduce our water consumption, which saves water costs and maintenance costs. We are increasing the access to daylight and views of the natural environment for our patients and staff, improving the quality of the work environment with little or no additional costs. On one project we are using a photovoltaic array to screen views of rooftop mechanical equipment. By taking advantage of state-sponsored energy credits, this system costs less than a conventional mechanical screen. We have also taken significant steps toward eliminating the use of PVC in building materials.

That is a quick overview of just a few of our efforts. Today I would like to focus on some of our materials and resources initiatives because they have direct health impacts on our staff, patients, and communities. And it is an area that would benefit from additional research.

This is a story that illustrates how we were able envision a future that is quite different from the present and ask for products that did not exist at the time. As large-scale consumers, we were able to create incentives to transform the market place. Kaiser Permanente's National Facilities Services (NFS) division manages the design, construction, and operation of all our buildings. NFS has a robust standards program to control quality, facilitate design, ensure operational efficiency, and promote our green buildings program. The national purchasing agreement

(NPA) program was established in 1991 and is an integral part of this effort. The NPA is comprised of 25 contracts with manufacturers of contractor-furnished and installed systems and materials. It includes a wide variety of items from lighting and HVAC equipment to flooring and ceiling tiles. The intent of the program is to partner with the manufacturers to realize the goals of our standards program while reducing our first and life cycle costs. Compliance with the NPAs is mandatory for our designers, and our strategic alliance allows us to help develop products and systems that meet our specific needs.

In 1993 Kaiser Permanente negotiated the first NPAs for carpet. We included in our request for proposal (RFP) a requirement that bidders state what they were doing to reduce waste and support recycling. We were not pleased with the responses we received. The responses either omitted recycling or included programs that sent carpet to road construction contractors for curing concrete, a one-time reuse, then it was thrown away. Only one company was actually recycling carpet. The rest did not comprehend why we were even asking the question. That one company, C&A, was successful in becoming part of the NPA along with two other companies. In the next nine years we dropped one of the three companies, continued to partner with C&A, and tried to work with the other company to enhance recycling and landfill diversion.

In 2002, when the NPA contracts for carpet came up for renewal, Kaiser Permanente decided to focus on sustainability in looking at our current and potential partners. Our negotiating team included interior designers, a representative from our environmental services (janitorial) division, as well as our director of environmental stewardship. We also included two other members of our green buildings committee: an outside architect and a representative from the Healthy Building Network. The team was charged with focusing on three main criteria in evaluating current and potential bidders: sustainability, product performance, and aesthetics.

The negotiating team conducted research into the carpet industry to identify which companies were truly leading the charge to sustainability. We also met with fiber manufactures to try to better understand the environmental impact of carpet fiber. After sorting out the facts from the "greenwashing," the team decided to look at five carpet companies, including the two under contract. The three other mills were included based on their leadership in the industry for sustainable practices.

The negotiating team then prepared an RFP, which was sent to all five companies. The Healthy Building Network helped us by developing a very detailed questionnaire that looked at the environmental impact of carpet from manufacturing to the end of its life and beyond. The RFP contained an extensive product performance questionnaire that included a requirement for impact test results for the backing. This is because we needed to determine if their backing was truly impermeable. They were also required to submit carpet samples of the products that they proposed for inclusion in our standards. Each company was then invited to make a presentation to the team that focused on sustainable practices, their healthcare product line, and product performance.

The team then met and scored each company based on the selection criteria. Sustainable issues were given 45 percent weight, product evaluation was given 45 percent weight, and green innovation was given a 10 percent weight.

As with Microsoft and Hewlett-Packard, a major issue for Kaiser Permanente is eliminating PVC from products because it contributes to dioxin pollution (Microsoft is curbing use of PVC, 2005). Based on our assessment of carpet in our existing facilities, there was no question that vinyl-backed carpet outperformed broadloom and had the advantage of potentially being recycled into new carpet at the end of use. Our hope was to find non-PVC backed carpet that would have similar performance characteristics to the vinyl-backed products we were using. However, none of the non-PVC-backed products passed the dynamic impact tests we required. So the team focused on what companies were doing in their research and development to create an alternative to PVC and how likely they were to partner with Kaiser Permanente in that quest.

Based on our analysis of the five companies, Kaiser Permanente did not renew the contract with one of the original two companies and added a new one. These two manufacturers were C&A and Interface. Both carpet manufactures were given two years to develop a non-PVC-backed carpet. We monitored each company's progress, pilot tested PVC-free carpets as they were developed, and reviewed the lab tests we required.

Last year C&A developed Ethos, a carpet with backing that has the same level of performance as PVC without the PVC. Ethos uses a backing material that is reclaimed from laminated safety glass. As a result, the backing has 96 percent postconsumer recycled content. Needless to say, our carpet NPA is now solely with C&A. The market has been transformed. Kaiser Permanente is paying the same amount for its carpeting, and Ethos is now available to other healthcare carpet consumers.

Where Do We Go from Here?

We have revised our standards to require the use of sheet flooring and tile flooring that do not contain PVC. Currently these products are not less expensive than the products they are replacing. However, as we explore the health impacts of these products and the products used to clean and maintain them we are finding other advantages. The alternative flooring products we are using, Stratica by Amtico and Nora rubber flooring, have a higher coefficient of friction, and early studies of facilities where they have been used are indicating a significant reduction in slip, trip, and fall injuries as compared to our facilities with vinyl flooring. Stratica and Nora rubber floors also do not require waxing and buffing, which results in lower maintenance costs. We believe eliminating waxing and buffing also results in less asthma-triggering particulates and harsh chemical fumes in our facilities, which should have a beneficial health impact on our patients, physicians, and staff. There already is some research and scientific literature to support these conclusions, but there certainly is room for more (Bornehag et al., 2004b).

Consumers, like us, would benefit from additional research on the health impacts of the products we use to build and maintain our facilities. We also would benefit from a product content labeling system that reveals the chemicals that are in the materials we use to build and furnish our facilities. This would enable consumers to make informed choices. It will help us fulfill our environmental mission and facilitate our ability to make informed decisions that will benefit our health and the health of generations to come.

References

AHA. 1998. *Memorandum of understanding between the American Hospital Association and the United States Environmental Protection Agency.* http://www.h2e-online.org/docs/h2emou101501.pdf (accessed January 30, 2007).

Alevantis, L., K. Frevert, R. Muller, H. Levin, and A. Sowell. 2002. Sustainable building practices in California state buildings. In *Proceeding of Indoor Air 2002: 9th International Conference on Indoor Air Quality and Climate* 3:666-671.

Alexeeff, G. V., J. D. Budroe, J. F. Collins, D. E. Dodge, J. R. Fowles, J. B. Faust, R. F. Lam, D. C. Lewis, M. A. Marty, F. J. Mycroft, H. Olson, J. S. Polakoff, J. Rabovsky, and A. G. Salmon. 2000. *Noncancer chronic ceference exposure levels.* Oakland: California Environmental Protection Agency, Office of Environmental Health Hazard Assessment, Air Toxicology and Epidemiology Section. http://www.oehha.ca.gov/air/chronic_rels/pdf/relsP32k.pdf (accessed May 3, 2007).

Alvarez, L. 2004. Where the healing touch starts. *New York Times.* September 7. http://www.nytimes.com/2004/09/07/health/07hosp.html?ex=1252296000&en=aefa2c6a90824d76&ei=5090&partner=rssuserland (accessed May 2, 2007).

Antonovsky, A. 1987. The salutogenic perspective: Toward a new view of health and illness. *Advances* 4:47–55.

ASHE (American Society for Healthcare Engineering). 2002. *Green Healthcare Construction Guidance Statement.* http://www.healthybuilding.net/healthcare/ASHE_Green_Healthcare_2002.pdf (accessed April 26, 2007).

Beauchemin, K. M., and P. Hays. 1998. Dying in the dark: Sunshine, gender and outcomes in myocardial infarction. *Journal of the Royal Society of Medicine* 91:352–354.

Beauchamp, T. L. *Principles of biomedical ethics*, 5th Edition. New York: Oxford University Press, 2001.

Benedetti, F., C. Colombo, B. Barbini, E. Campori, and E. Smeraldi. 2001. Morning sunlight reduces length of hospitalization in bipolar depression. *Journal of Affective Disorders* 62:221–223.

Bergsland, K. H. 2005. *Nytt Rikshospital tre år etter – hva har bygget betydd for driften?* (Nytt Rikshospital three years after—how does the new facility influence the running of the hospital?). Trondheim, Norway: Stiftelsen for industriell og teknisk forskning ved NTH (SINTEF).

Bernheim, A., and H. Levin. 1997. Material selection for the public library. *Proceedings of Healthy Buildings/IAQ 1997* 3:599–604.

Bernheim, A., H. Levin, and L. Alevantis. 2002. Special environmental requirements for a California state office building. *Proceeding of Indoor Air 2002: 9th International Conference on Indoor Air Quality and Climate* 4:918–923.

Berry, L. 1999. *Discovering the soul of service: The nine drivers of sustainable business success.* New York: The Free Press.

Berry L. L., D. Parker; R. C. Coile, Jr., D, K. Hamilton, D. D. O'Neill, and B. L. Sadler. 2004. Can better buildings improve care and increase your financial returns? *Frontiers of Health Services Management* 21:3–24.

Biley, F. C. 1994. Effects of noise in hospitals. *British Journal of Nursing* 10-23(3):110–113.

Bjorseth, O., J. Bakke, N. Iversen, and B. Martens. 2002. *Characterization of emissions from mechanical polishing of PVC floors.* http://www.nyf.no/bergen2002/papers/ abstracts/L-abstr.pdf (accessed January 31, 2007).

Black, M., W. Pearson, J. Brown, and S. Sadie. 1993. Material selection for controlling IAQ in new construction. *Proceedings of Indoor Air* 2:611–616.

Blomkvist, V., C. A. Eriksen, T. Theorell, R. Ulrich, and G. Rasmanis. 2005. Acoustics and psychosocial environment in intensive coronary care. *Occupational and Environmental Medicine* 62:e1.

Bornehag, C., J. Sundell, and T. Sigsgaard. 2004a. Dampness in buildings and health (DBH): Report from an ongoing epidemiologic investigation on the association between indoor environmental factors and health effects among children in Sweden. *Indoor Air* 14:59–66.

Bornehag, C. G., J. Sundell, C. J. Weschler, T. Sigssgaard, B. Lundgren, M. Hasselgren, and L. Hagerhed-Engman. 2004b. The association between asthma and allergic symptoms in children and phthalates in house dust: A nested case-control study. *Environmental Health Perspectives* 112:1393–1397.

Boyden, S. 1971. Biological determinants of optimal health. In the *Human Biology of Environmental Change Conference.* April, Blantyre, Malawi, 5–12.

Boyden, S. 2004. *The biology of civilization: Understanding human culture as a force in nature.* Sydney, New South Wales: University of New South Wales Press.

Brent RJ. *Cost-benefit analysis and health care evaluations.* Northampton: Edward Elgar Publishing, 2003.

Cabrera, I. N., and M. H. Lee. 2000. Reducing noise pollution in the hospital setting by establishing a department of sound: A survey of recent research on the effects of noise and music in health care. *Preventive Medicine* 30:339–345.

Carnegie Mellon University Center for Building Performance. 2005 (unpublished). From the ongoing *Building Investment Decision Support (BIDS) research project with funding from the Advanced Building Systems Integration Consortium (ABSIC)*; contact: Vivian Loftness. Pittsburgh, PA.

Christensson, B., A. Jansson, and J. Johansson. 2002. *Particles indoor in sick and healthy buildings.* IVL Swedish Environmental Research Institute Ltd. http://www.nyf.no/bergen2002/papers/abstracts/L-abstr.doc (accessed January 31, 2007).

CDC (Center for Disease Control and Prevention). 2000. *Hospital infections cost U.S. billions of dollars annually.* http://www.cdc.gov/drugresistance/healthcare/problem.htm (accessed March 17, 2007).

Chow S-C., and J.P. Liu. *Design and analysis of clinical trials: Concepts and methodologies.* Indianapolis, IN: Wiley Interscience, 2004.

CIA (Central Intelligence Agency). 2007. *World fact book 2007.* Washington, DC: CIA Library

CIWMB (California Integrated Waste Management Board). 2007. *Green project specifications.* http://www.ciwmb.ca.gov/greenbuilding/specs/ (accessed March 16, 2007).

CIWMB (California Integrated Waste Management Board). 2000. *Green building design and construction guidelines.* http://www.ciwmb.ca.gov/GreenBuilding/ (accessed January 30, 2007).

Conlin, M. 2000. Is your office killing you? *Business Week.* http://www.businessweek.com/ 2000/00_23/b3684001.htm (accessed January 30, 2007).

CRI (Carpet and Rug Institute). 2004. Carpet and Rug Institute launches Green Label Plus Program. *The Carpet and Rug Institute.* www.carpet-rug.org/News/040614_GLP.cfm (accessed May 2, 2007).

Dean, L. 2000. Conference Remarks. In Conference summary *Setting healthcare's environmental agenda October 16.* http://www.noharm.org/details.cfm?type=document&ID=477 (accessed January 31, 2007).

Dick Corporation. 2006. Childrens hospital project to be a local gem. *A Publication of Dick Corporation.* http://www.dickcorp.com/dickcorp/dickbuilds/default.asp?vol=22&num=2&art=109 (accessed March 1, 2007).

Donaldson C, M. Mugford, L. Vale, Eds. *Evidence-based health economics.* London, UK: BMJ Books, 2002.

Douglas, M. R., and C. H. Douglas. 2004. Patient-friendly hospital environments: Exploring the patients' perspective. *Health Expectations* 7:61–73.

Drummond, M., B. O'Brien, G. Stoddart, G. Torrance. *Methods for the economic evaluation of health care programmes,* 2nd edition. New York: Oxford University Press, 1997.

Eastman C. I., M. A. Young, L. F. Fogg, L. Liu, and P. M. Meaden. 1998. Bright light treatment of winter depression: A placebo-controlled trial. *Archives of General Psychiatry* 55:883–889.

EIA (Energy Information Administration). 1995. *Commercial Buildings Energy Consumption Survey.* http://www.eia.doe.gov/emeu/consumptionbriefs/cbecs/pbawebsite/health/health_howuseelec.htm (accessed 5/03/07).

Elkington J. 1998. *Cannibals with forks: The triple bottom line of 21st century business.* Gabriola Island, BC: New Society Publishers.

Engelhardt H.T. *The foundations of bioethics,* 2nd Edition. New York: Oxford University Press, 1995.

EPA (Environmental Protection Agency). 1996. *National dioxin emissions from medical waste incinerators.* http://www.epa.gov/ttnatw01/129/hmiwi/iv-a-7.pdf (accessed March 1, 2007).

EPA. 2006. *Why study indoor air quality.* http://www.epa.gov/region01/eco/iaq/whyiaq.html (accessed January 31, 2007).

Esty, D. C., and A. S. Winston. 2006. *Green to gold: How smart companies use environmental strategy to innovate, create value, and build competitive advantage.* New Haven, CT: Yale University Press.

Friedman, L. M., C. D. Furberg, and D. L. DeMets. *Fundamentals of clinical trials,* Third edition. New York: Springer, 1999.

Fung, F., and W. Hughson. 2003. Health effects of indoor fungal bioaerosol exposure. *Applied Occupational and Environmental Hygiene* 18:535–544.

Furtaw, E., M. Pandian, D. Nelson, and J. Behar. 1996. Modeling indoor air concentrations near emission sources in imperfectly mixed rooms. *Journal of Air and Waste Management Association* 46:861–868.

Gawande A. 2007. *Better: A surgeon's notes on performance.* New York: Metropolitan Books.

GGHC (Green Guide for Health Care). 2006. *The green guide for health care, version 2.2.* http://www.gghc.org (accessed January 30, 2007).

Hawthorne, C. 2003. Turning down the global thermostat. *Metropolis Magazine.* http://www.metropolismag.com/html/content_1003/glo/index.html (accessed January 30, 2007).

Hendrich, A., A. Nyhuis, T. Kippenbrock, and M. E. Soja. 1995. Hospital falls: Development of a predictive model for clinical practice. *Applied Nursing Research* 8:129–139.

Hill Burton Act. 1946. Public Law 79-725.

Hudnell, H. K., D. A. Otto, D. E. House, and L. Molhave. 1992. Exposure of humans to a volatile organic mixture, II, sensory—A selection of papers from Indoor Air '90 concerning health effects associated with indoor air contaminants. *Archives of Environmental Health* 47:31–38.

IOM (Institute of Medicine). 1999. *Crossing the quality chasm: A new health system for the 21st century.* Washington, DC: National Academy Press.

IOM. 2000. *To err is human: Building a safer health system.* Washington, DC: National Academy Press.

Jaakkola, J., L. Oie, P. Nafstad, G. Botten, S. Samuelsen, and P. Magnus. 1999. Interior surface materials in the home and the development of bronchial obstruction in young children in Oslo, Norway. *American Journal of Public Health* 89:188–192.

Jaakkola, J., H. Parise, V. Kislitsin, N. Lebedeva, and J. Spengler. 2004. Asthma, wheezing, and allergies in Russian schoolchildren in relation to new surface materials in the home. *American Journal of Public Health* 94:560–562.

Jackson, R., and H. Frumkin. 2004. Urban sprawl and public health. *Public Health Reports* 117: 201–217.

Johnsen, R. 2006. Health systems in transitions. *European Health Systems Observatory* 8(1):31. http://www.euro.who.int/Document/E88821.pdf (accessed March 9, 2007).

Katz, D. L. 2006. *Clinical epidemiology & evidence-based medicine: Fundamental principles of clinical reasoning & research*. Thousand Oaks, CA: Sage Publications.

Kats, G. 2003. *Green building costs and financial benefits*. Massachusetts Technology Collaborative.

Kats, G. 2006. *Greening America's schools costs and benefits*. Capitol-E.com. Washington, DC: Capitol E Publications. http://www.cap-e.com/ewebeditpro/items/O59F9819.pdf (accessed March 9, 2007).

Kats, G., and Capital E. 2003. *The costs and financial benefits of greening buildings*. A report to California's sustainable building task force October 2003.

Kozlowski, D. 2004. Green building report: Guides make it easier to green hospitals. *Building Operating Management* November. http://www.facilitiesnet.com/bom/article.asp?id=2265&keywords= (accessed May 3, 2007).

Lawrence, D. 2000. Conference remarks. In Conference summary *Setting healthcare's environmental agenda October 16*. http://www.noharm.org/details.cfm?type= document&ID=477 (accessed January 30, 2007).

Lent, T. 2006. *Improving indoor air quality with the California 01350 specification*. Healthy Building Network. http://www.healthybuilding.net/healthcare/CHPS_1350_summary.pdf #search='CRE Ls%2080%20chemicals' (accessed January 31, 2007).

Lerner, M. 2000. Conference remarks. In Conference summary *Setting healthcare's environmental agenda October 16*. http://www.noharm.org/details.cfm?type= document&ID=477 (accessed January 30, 2007).

Levin, H. 1998. Toxicology-based air quality guidelines for substances in indoor air. *International Journal of Indoor Air Quality and Climate* 5:5–7.

McDonough, W. 1998. The next industrial revolution. *The Atlantic Monthly* 282(4):82–92.

McLennan, J. F. 2004. *The philosophy of sustainable design*. Bainbridge Island, WA: Ecotone Publishing Co.

Microsoft is curbing use of PVC, a popular plastic. 2005. *Wall Street Journal,* December 7, D7.

Norbäck, D., G. Wieslander, K. Nordstrom, and R. Walinder. 2000. Asthma symptoms in relation to measured building dampness in upper concrete floor construction, and 2-3ethyl-1-hexanol in indoor air. *International Journal of Tuberculosis and Lung Disease* 4:1016–1025.

NRC (National Research Council). 2002. *Multiple chemical sensitivities*. Washington, DC: The National Academy Press.

NRC. 2006. *Green schools: Attributes for health and learning*. Washington, DC: The National Academies Press.

OFEE (Office of Federal Environmental Executive). 2003. *Federal commitment to green building: Experiences and expectations*. http://www.ofee.gov/sb/fgb_report.html (accessed January 30, 2007).

Oie, L., P. Nafstad, G. Botten, P. Magnus, and J. Jaakkola. 1999. Ventilation in homes and bronchial obstruction in young children. *Epidemiology* 10:294–299.

Orians, G. H., and J. H. Heerwagen. 1992. Evolved responses to landscapes. In *The Adapted Mind*. 555–579.

Pierce, J., and A. Jameton. 2004. *The ethics of environmentally responsible health care*. New York: Oxford University Press.

Pommer, L., J. Fick, J. Sundell, C. Nilsson, M. Sjostrom, B. Stenberg, and B. Andersson. 2004. Class separation of buildings with high and low prevalence of SBS by principal component analysis. *Indoor Air* 14:16–23.

Progress report on sustainability. 2003. *Building Design & Construction* November:22–24.

RICS (Royal Institution of Chartered Surveyors). 2005. *Green value, green buildings, growing assets.* http://www.gvrd.bc.ca/buildsmart/pdfs/greenvaluereport.pdf (accessed February 02, 2007).

Roberts, M. J., and M. R. Reich. 2002. Ethical analysis in public health. *Lancet.* 359:1055–1059.

Roshanaei, H., and D. Braaten. 1996. Indoor sources of airborne particulate matter in a museum and its impact on works of art. *Journal of Aerosol Science* 27:443–444.

Rubin, H. R., A. J. Owens, and G. Golden. 1998. *Status report: An investigation to determine whether the built environment affects patient medical outcomes.* Martinez, CA: Center for Health Design.

Sadler, B. L. 2004. Designing with health in mind; Innovative design elements can make hospitals safer, more healing places. *Modern Health Care* 34:28.

Savitz, A. W., and K. Weber. 2006. *The triple bottom line: How today's best-run companies are achieving economic, social and environmental success—and how you can too.* San Francisco, CA: Jossey-Bass.

Schickler, P. 2005. Achieving health or achieving well being? *Learning in Health and Social Care* 3:217–224.

Schettler, T. 2001. *Environmental challenges and visions of sustainable health care.* Presented at CleanMed Conference, Boston. http://www.sehn.org (accessed January 30, 2007).

SCS (Scientific Certification System). 2005. *Floor score.* http://scscertified.com/iaq/floorscore.html (accessed April 26, 2007).

Seligman, M. E., and M. Csikszentmihalyi. 2000. Positive psychology: An introduction. *American Journal of Psychology* 55:5–14.

Shertzer, K. E., and J. F. Keck. 2001. Music and the PACU environment. *Journal of Perianesthesia Nursing* 16:90–102.

Skyberg, K., K. Skulberg, W. Eduard, E. Skaret, F. Levy, and H. Kjuus. 2003. Symptoms prevalence among office employees and associations to building characteristics. *Indoor Air* 13:246–252.

Slotterback, S. 2006. *Building green at a large scale.* PowerPoint Presented at the IOM Green Healthcare Institutions: Health, Environment, and Economics Workshop, Washington, DC. http://www.iom.edu/?id=33245 (accessed February 22, 2006).

Smith P.C. 2005 Performance measurement in health care: History, challenges and prospects. *Public Money & Management* 25:213–220.

Spengler, J., and Q. Chen. 2000. Indoor air quality factors in designing a healthy building. *Annual Review of Energy and the Environment* 25:567–601.

Sundell, J. 1999. *Indoor environment and health.* Stockholm, Sweden: National Institute of Public Health.

Sustainability Commitee. 2005. *Sustainability vision for Emory.* Atlanta, Georgia: Emory University. http://www.finadmin.emory.edu/policies/SustyReportFinal.pdf (accessed May 3, 2007).

Sustainable Development Commission. 2006. *Healthy Futures* 4. http://www.sd-commission.org.uk/publications/downloads/HF4-final.pdf (accessed April 26, 2007).

Thomas, J. C., M. Sage, J. Dillenberg, V. J. Guillory. 2002. A code of ethics for public health. *American Journal of Public Health* 92:1057–1059.

Topf, M. 1992. Stress effects of personal control over hospital noise. *Behavioral Medicine* 18:84–94.

Tuomainen, A., M. Seuri, and A. Sieppi. 2004. Indoor air quality and health problems associated with damp floor coverings. *International Archives of Occupational and Environmental Health* 77:222–226.

Turner Green Building. 2006. *Turner green building market barometer.* http://www.turnerconstruction.com/corporate/files_corporate/Green_Survey.pdf (accessed February 05, 2007).

Ulrich, R. 1991. Effects of healthcare interior design on wellness: Theory and recent scientific re-search. *Journal of Healthcare Interior Design* 3:97–109.

Ulrich, R. S. 1999. Effects of gardens on health outcomes: Theory and research. *Healing gardens*, edited by C. Cooper, C. C. Marcus, and M. Barnes. New York: Wiley. Pp. 27–86.

Ulrich, R. S., C. Zimring, X. Quan, A. Joseph, and R. Choudhary. 2004. *The role of the physical environment in hospitals of the 21st century: A once in a lifetime opportunity.* Princeton, NJ: Robert Wood Johnson Foundation.

USGBC (U.S. Green Building Council). 2006. Leadership in energy and environmental design. http://www.usgbc.org/DisplayPage.aspx?CategoryID=19 (accessed January 31, 2007).

Walch, J. M., B. S. Rabin, R. Day, J. N. Williams, K. Choi, and J. D. Kang. 2005. The effect of sunlight on postoperative analgesic medication use: A prospective study of patients undergoing spinal surgery. *Psychosomatic Medicine* 67:156–163.

Willard, B. 2002. *The sustainability advantage: Seven business case benefits of a triple bottom line.* Gabriola Island, BC: New Society Publishers.

Wolkoff, P., P. Clausen, B. Jensen, G. Nielsen, and C. Wildins. 1997. Are we measuring the relevant indoor pollutants? *Indoor Air* 7:92–106.

Appendix A

Workshop Agenda

GREEN HEALTHCARE INSTITUTIONS:
HEALTH, ENVIRONMENT, AND ECONOMICS

Sponsored by
the Roundtable of Environmental Health Sciences, Research, and Medicine

January 10, 2006

8:30 a.m.	**Welcome and Opening Remarks** Paul G. Rogers, J.D. Roundtable chair
8:40 a.m.	**Remarks and Charge to Participants** Howard Frumkin, M.D., Dr.P.H. Roundtable member
8:55 a.m.	**Green Building and Health Agendas: Points of** **Convergence** Craig Zimring, Ph.D. Professor of Architecture Georgia Institute of Technology

SESSION I: CURING THE "GREEN" WAY: STORIES OF SUSTAINABLE HEALTHCARE FACILITIES

Scope: Currently, there are only a select number of case studies of health-care facilities that are being built "green." This session sets the stage for further discussion by examining two case studies in order to understand the decision-making process involved, the research data to support the case, and the health outcomes (if monitored).

Moderator:	Barbara M. Alving, M.D., Acting Director of National Center for Research Resources (NCRR)

9:15 a.m.   **The Green Guide for Health Care—A Tool for High-Performance Healing Environments**
                      Robin Guenther, AIA
                      Architect
                      Guenther 5 Architects

9:35 a.m.   **Building Green and Integrating Nature: Rikshospitalet University Hospital in Oslo, Norway**
                      Knut Bergsland, AIA
                      Senior Advisor
                      Hospital Planning Department
                      SINTEF Health Research

10:00 a.m.   **General Discussion**

10:20 a.m.   **Break**

SESSION II: THE CASE FOR GREEN BUILDINGS PART I: ECONOMICS, ETHICS, AND EMPLOYMENT

Scope: This session will consider some of the benefits for building green by looking at research data in economics and the social sciences. The main question centers around how industry and society value buildings both in the short- and long-term.

Moderator: Alexis Karolides, M.Arch. Roundtable member

10:50 a.m.   **The Financial Implications of Health and Productivity Gains in Green Commercial, Public, and Education Buildings**
                      Gregory Kats, M.B.A.
                      Principal
                      Capital E

11:20 a.m.   **Building Green and Ethics**
                      John Poretto, B.S.
                      President
                      Sustainable Business Solutions

11:50 a.m.   **Staff Retention: Can a Healthy Environment Make a Difference?**
George Bandy II
Sustainable Programs Manager
Interface, Inc.

12:20 p.m.   **General Discussion**

12:45 p.m.   **Lunch**

SESSION III: THE CASE FOR GREEN BUILDINGS II: HEALTH

Scope: This session will look at the current state of the knowledge of the linkages between building green and human health. Through presentations and discussions, the roundtable will gain a better understanding of the state of the research and the research gaps.

Moderator: Henry Hatch, Lt General (Ret), Chief of Engineers and Commander of the U.S. Army Corps of Engineers, and Chair, NRC Board on Infrastructure and the Constructed Environment and the Federal Facilities Council

2:00 p.m.   **Design Principles in Healthy Building**
Anthony Bernheim, FAIA, LEED AP
Principal, Green Design, SMWM Architecture Planning + Urban Design

2:20 p.m.   **Building-Related Health Effects: What Do We Know?**
Ted Schettler, M.D., M.P.H.
Science Director
Science and Environmental Health Network

2:50 p.m.   **The Relationship Between Environmental Design and Patient Medical Outcomes**
Judy Heerwagen, Ph.D.
Environmental Psychologist
J.H. Heerwagen & Associates, Inc.

3:15 p.m.   **The Fable Hospital: A Business Case for Better Building**
Derek Parker, FAIA, RIBA, FACHA
Chairman
Anshen and Allen Architects, Inc.

3: 45 p.m.   **Discussion**

4:10 p.m.   **Break**

SESSION IV: RECOMMENDATIONS FOR FUTURE RESEARCH: WHAT DO WE NEED TO KNOW?

Scope: Continuation of the discussion from sessions II and III by discussing the research needs for the area of building green and human health.

Moderator: Russell Perry, AIA, LEED AP, Principal, SmithGroup

4:30 p.m.   **General Discussion**

5:30 p.m.   **Adjourn**

January 11, 2006

8:30 a.m.   **Welcome Back**

SESSION V: THE PROCESS OF CHANGE

Moderator: Nancy L. Hughes, RN, M.H.A., Director, Center for Occupational and Environmental Health, American Nurses Association

8:55 a.m.   **Framing the Process: Institutional Change to Greening a Campus**
Bahar Armaghani, B.S., LEED AP
Project Manager/Quality Assurance Coordinator
Facilities Planning and Construction
University of Florida

9:25 a.m.   **The Conflict Between Growth and Going Green: The Experience at Emory**
Wayne Alexander, M.D., Ph.D.
R. Bruce Logue Professor and Chair
Emory University School of Medicine

9:40 a.m.   **General Discussion**

10:05 a.m.   **Break**

SESSION VI: CHAMPIONS FOR CHANGE

Moderator: Roger Bulger, M.D., Roundtable member

10:35 a.m.   **Implementing the Decision to Build Green in University Medical Centers**
         Roger Oxendale, M.B.A.
         President and CEO
         Children's Hospital of Pittsburgh

11:05 a.m.   **Building Green on a Large Scale**
         Scott Slotterback, M.S.
         Senior Project Manager
         Kaiser Permanente National Facilities Services Project
             Administration Group

11:35 a.m.   **General Discussion**

12:05 p.m.   **Final Summation**
         Howard Frumkin, M.D., Dr.P.H.
         Roundtable member

12:20 p.m.   **Adjourn**

Appendix B

Speakers and Panelists

Wayne Alexander
R. Bruce Logue Professor and Chair
Emory University School of Medicine

Barbara Alving
Acting Director
National Institutes of Health

Bahar Armaghani
Project Manager
University of Florida

George Bandy II
Sustainable Programs Manager
Interface, Inc.

Knut Bergsland
Senior Advisor
SINTEF Health Research

Anthony Bernheim
Managing Principal
SMWM Corporation

Robin Guenther
Architect
Guenther 5 Architects

Henry Hatch
Lt. General (Ret), Chief of
    Engineers, and Chair, NRC
    Board on Infrastructure and the
    Constructed Environment

Judith Heerwagen
J.H. Heerwagen & Associates, Inc.

Nancy Hughes
Director
Center for Occupational and
    Environmental Health

Gregory H. Kats
Principal
Capital E

Roger Oxendale
President and CEO
Children's Hospital of Pittsburgh

Derek Parker
Chairman
Anshen and Allen Architects, Inc.

Russell Perry
Principal
SmithGroup

Ted Schettler
Science Director
Science and Environmental Health
    Network

Scott Slotterback
Senior Project Manager
Kaiser Permanente National Facilities
    Services

Craig Zimring
Professor of Architecture
Georgia Institute of Technology

Appendix C

Workshop Participants

Stephen Ashkin
The Ashkin Group, LLC

Meredith Banasiak
Academy of Neuroscience for
 Architecture

Marcia Barr
Center for Environmental Oncology

Sheila Bosch

William Brodt
National Aeronautics and Space
 Administration

Carl Brooks
Emergency Care Research Institute

Sarah Buchwalter
ICF Consulting/Energy Star

Nancy Bullock
Elizabeth Seton Pediatric Center

Orest Burdiak
Department of Veterans Affairs

Paula Burgess
National Center for Environmental
 Health/Agency for Toxic
 Substances and Disease Registry/
 Center for Disease Control and
 Prevention

Beata Canby
Elizabeth Seton Pediatric Center

Amy Carpenter
Wallace, Roberts & Todd

Margaret Chu
U.S. Environmental Protection
 Agency

Maryann Donovan
University of Pittsburgh Cancer
 Institute

Glen Dorsey
Heinz Endowments

Clarence Dukes
National Institute of Health

M. Michael DunGan
Naval Facilities Engineering
    Command HQ

John Eberhard
Academy of Neuroscience for
    Architecture

Donald Emmerling
U.S. Army Corps of Engineers

Margaret Fowke
National Weather Service

Tom Gaulke
Indian Health Service

Leslie Getzing
American Federation of Teachers

Anna Gilmore Hall
Health Care Without Harm

Yun Gu
Center for Building Performance and
    Diagnostics

John Hamilton
Testing Adjusting and Balancing
    Bureau

Winifred J. Hamilton
Baylor College of Medicine

Jeff Hardin
U.S. Army Corps of Engineers

Eric Haukdal
U.S. Department of Health and
    Human Services

Brandon Karlow
Institute of Medicine

Woodie Kessel
American Astronomical Society

Steven Knippen
Teknion

Joanne Krause
Medical Facilities Design Office

John Longstaff
Indian Health Service

Marcia Marks

Ellen Mazo

Leyla McCurdy
National Environmental Education &
    Training Foundation

Farhad Memarzadeh
National Institute of Health

Bob Musil
Physicians for Social Responsibility

Meghan Newcomer
President's Cancer Panel

Aimee O'Grady

Cheryl Phillips
Naval Facilities Engineering
    Command

Bradley Provancha
Washington Headquarters Services

Edward Rau
National Institute of Health

Bill Ravanesi
Health Care Without Harm

Steven Raynor
Indian Health Service

Clark Reed
U.S. Environmental Protection
    Agency

Katherine Seikel
Office of Pesticide Programs

Lloyd Siegel
Department of Veterans Affairs

Janice Simmons
The Quality Letter for Healthcare
    Leaders

Carl Smith
GREENGUARD Environmental
    Institute

Megan Snyder
Carnegie Mellon University

Lynda Stanley
National Research Council

Kathy Sykes
U.S. Environmental Protection
    Agency

Leonard Taylor
University of Maryland Medical
    Center

Marcella Thompson
Tetra Tech EM Inc.

Ward Thompson
HKS Architects

Esmail Torkashv
National Cancer Research Foundation

Gail Vittori
Center for Maximum Potential
    Building Systems

Carol Walker
University of Florida

Calvin Williams
National Aeronautics and Space
    Administration

Joanna Winchester
Sierra Club, Washington, DC, chapter

Dan Winters
Evolution Partners

Andrew Wolman

Ariel Wyckoff
U.S. Environmental Protection
    Agency

Ariel Wyckoff
Internal Revenue Services Facilities

Richard Zdanis
MD Hospitals for a Healthy
    Environment